Introduction to Block 3

Success in developing and exploiting a piece of engineering invention, or indeed any product in the marketplace, is based on the ability to communicate: communication of the concept, communication of how to make it, and communication of how the product conforms with standards or legislation.

This may sound complex, but you will certainly have experienced this yourself in one way or another. If you go to a DIY shop to buy nails of a certain size, you need to know what size is required, and the shop needs to have the product suitably labelled. If you want to buy fabric or wallpaper, you need to be sure that your measurements will produce the right amount of material when you give them to the dealer.

You have probably been involved in the specification for an engineering project at some time, whether or not you realized it. It may have been as simple as deciding the layout for a new kitchen, or working with an architect and builders to produce an extension to a house. In this case, the builder had to be able to convert the architect's drawings into the final construction (and produce a building which is structurally sound). There are also regulations which must be complied with. For example, a garage for a car must have a slightly sloping floor, so that in case of a fuel spillage the petrol will run away from the house. A buildings inspector would not approve an extension or redevelopment that did not comply.

In this block of the course you will study several examples of how engineers communicate as part of their activity. Part 1 looks at engineering standards. These are mechanisms by which engineers in different companies or in different countries achieve uniformity between their approach to a particular aspect of their work. It is an area where quite firm definitions are often made. Part 2 looks at an area where the wording of definitions can become critical, which is in the construction of patents to protect the ideas of an inventor or development team. In Part 3, we look at an area where definitions become highly imprecise, but which is none the less becoming one of the most important considerations for engineers in many fields: risk assessment. In Part 4 we shall look at some of the fundamental natural laws which constrain what engineering can achieve. Finally, in Part 5 we take an example of how changes in technology can be driven by changes in legislation: a case of environmental legislation forcing development of new materials and processes to replace an existing technique.

T173
Technology: A Level 1 Course
Engineering the Future

Block 3
Engineerir
Parts 1 an

The Open University

T173 Course Team

The following course-team members were responsible for this block.

Academic Staff

Dr Michael Fitzpatrick (Course Team Chair)
Dr Rod Barratt
Professor Nicholas Braithwaite
Professor Lyndon Edwards
Dr Peter Lewis
James Moffatt
Professor Bill Plumbridge
Dr George Weidmann
Adrian Demaid
Mark Endean
Jim Flood
Dr Bill Kennedy
Dr Suresh Nesaratnam

Consultant

Ted Bloomfield

Production Staff

Sylvan Bentley (Picture Research)
Philippa Broadbent (Materials Procurement)
Daphne Cross (Materials Procurement)
Tony Duggan (Project Control)
Andy Harding (Course Manager)
Richard Hoyle (Designer)
Allan Jones (Editor)
Lara Mynors (Editor)
Howie Twiner (Graphic Artist)
Patricia Telford (Course Team Secretary)

The Open University
Walton Hall, Milton Keynes
MK7 6AA

First published 2001. Second edition 2002.

Edited, designed and typeset by The Open University.

Printed and bound in the United Kingdom by The Burlington Press (Cambridge) Ltd.

ISBN 0 7492 8459 5

This text forms part of an Open University course T173 *Engineering the Future*. Details of this and other Open University courses can be obtained from the Call Centre, PO Box 724, The Open University, Milton Keynes MK7 6ZS, United Kingdom: tel. +44 (0)1908 653231.

For availability of this or other course components, contact Open University Worldwide Ltd, The Berrill Building, Walton Hall, Milton Keynes MK7 6AA, United Kingdom: tel. +44 (0)1908 858585, fax +44 (0)1908 858787, e-mail ouwenq@open.ac.uk

Alternatively, much useful course information can be obtained from the Open University's website http://www.open.ac.uk

2.1

Part 1
Engineering standards

Contents

1 Introduction

You should recall from Block 1 that the mass production of musket locks for the French army required uniformity in dimensions of all the components involved, so that they were completely interchangeable. We have come a long way since then in our ability to specify and manufacture items that are completely interchangeable no matter where they are made. An important aspect of this is the process known as standardization, which is one of the keys to successful engineering on any scale.

Establishing uniformity can take many forms, and you will have met examples of this in your own experience. Whenever you buy a new filter for your vacuum cleaner, or a box of floppy disks for your computer, or a refill for your ballpoint pen, you expect that the product will fit the item for which it was bought, even when it is made by a different manufacturer. If the disk did not fit into your computer, or the ink refill leaked badly because of a poor fit, you would probably return to the retailer and ask for a refund. Eventually the manufacturer would have to reimburse the cost of the faulty goods, and make a loss on those items.

Each of the examples that I have described, and I'm sure that you can think of many others, are manufactured to a particular specification. This specification would have designated the physical dimensions of the items, and the allowable tolerances: the removable disk has to be the right size to fit into the slot in the computer, and it must have correctly dimensioned slots in the centre to allow the drive mechanism to engage properly. Furthermore, all manufacturers of disks have to get it right: few purchasers are particularly choosy about brand names for something as common as computer disks, and they expect the disks to work regardless of where they were bought or by whom they were made. To complicate matters further, there are many different types of storage disk or cartridge for computer data (Figure 1.1), from the CD-ROM to the floppy disk to DAT tape drives. Each has its own particular characteristics and associated specifications.

Figure 1.1 Different computer data storage media, of all shapes and sizes!

Another common example of uniformity is the voltage used for mains electricity throughout Europe. It is nominally 230 volts a.c. Although plug designs differ, the electrical appliances themselves will usually work equally well in most European countries because of the uniformity of the power supply. This reduces costs for the manufacturers and, we are assured, for the consumer. Uniformity, then, can have advantages.

What, then, do the uniform dimensional requirements of musket locks, the specification of the computer disk, and the European harmonization of mains electricity all have in common?

In its broadest sense, they can all be regarded as examples of standardization, as indeed is the very system of units in which the dimensions or voltage are specified. There is a wide range of degrees of standardization, from the single manufacturer establishing uniform dimensions for a particular product such as the musket locks, through an industry agreeing a common set of dimensions for articles like computer disks or ballpoint pen refills, to national or international agreement on a common specification for something like mains electricity.

However, standardization is more usually regarded as that which is enshrined in documents called 'standards'. These may embody a particular method (for example, for manufacturing or testing) or concept, or suggest minimum requirements for a product. The thickness of the steel for an ammonia pressure vessel would typically be covered by a standard. Standards can take a variety of forms, such as descriptions of test techniques, information about products, guidelines about environmental responsibility, and so on. Some exist just to offer guidance for practitioners in a certain area of engineering; some are legally enforceable, and manufacturers ignore them at their peril. Other standards may be used as a basis for specifying a contract (for example, a contract may have a clause saying: 'The product must conform to Standard BS *wxyz*').

This section will investigate standards: their purpose, their derivation, their use and their limitations.

2 Standards and specifications

2.1 Why have standards?

You will recall from Block 1 that there are standard definitions for all the fundamental units of measurement: the metre for length; the kilogram for mass; the second for time. This is the simplest form of standard; that is, a statement of 'this is how it is' (for example, how something is defined, or how it is measured). A grocer is in trouble if the 5 kg of potatoes that you bought only read 3 kg on the scales when you get them home (assuming, of course, that it is not your weighing equipment that is at fault).

Frequently the purpose of a standard is to ensure that everyone is talking the same language. For example, many engineering standards concern definitions of the terms used in particular fields. Problems are certain when terms are misinterpreted (as with the ▼Mars Climate Orbiter▲). You want to be sure that 5 kg means the same to the grocer as it does to you. Engineers specifying the strength required of the steel cable for a road bridge want to know that they are talking the same language as the supplier of the material.

You can then take this argument a step further. The engineer will want to be assured that the test method used to obtain the stated value of strength is a sound, reproducible one which is accepted to give the 'right' answer. (I say 'right', using quotation marks, because for some properties there may be more than one way of obtaining a number to put against the value, but it is likely that only one of them will be the accepted standard route.) This may require any machines used in the testing procedure to be calibrated to a certain level of accuracy, which would again be specified by some form of standard.

This may sound like a horribly circular process, but in many cases careful calibration and standardization at many levels are needed in order to convince a customer that a product meets the required specification. Potentially this can save the customer money by removing the need for acceptance tests when the product arrives from the supplier (and also removing the need for the customer to maintain and calibrate test equipment of their own). There is now a series of standards designed to ensure the quality of product passed to a customer. Indeed a 'product' need not be a physical object; it could be a service, or a procedure, or almost anything else for which standardization might be considered desirable. We will take a look at an example of such standards later.

Standards also provide some reassurance to the customer. The power supply for the computer I am using to write this has a 'CE' mark on it, indicating that

▼Mars Climate Orbiter▲

A rather embarrassing failure to standardize occurred in 1999. A planetary probe which was supposed to go into orbit around Mars steered into the planet's atmosphere and burned up instead. The reason? The team responsible for generating software to drive the motors was working in the Imperial system of units, which measures force in pounds-force, lbf. However, other teams on the project expected that the output from the software would be in newtons (recall this unit from Block 1). The difference between the two systems of measurement is about a factor of 4.5, and hence 10 N is equivalent to about 2.2 lb f. So the numbers would have *looked* about right.

You might think that a factor of nearly five is quite large: certainly you would be surprised if your supermarket shopping bill was five times more or less than what you expected. However, the calculations of precisely how much thrust is required from a motor in order to send a spacecraft into orbit around a planet are highly complicated, involving the mass of the craft, its velocity (remember this is a vector, and so includes both speed and direction) and the gravitational force being exerted on it by the planet. Small wonder that it's unlikely that you would notice if the end result was out by a factor of five. A factor of twenty or a hundred might look wrong instantly, though.

it conforms to a European standard, in this case to prevent it interfering with other electrical equipment, or being unduly affected itself by equipment nearby. Many electrical appliances carry this mark (Figure 1.2) – in fact it is illegal for electrical equipment to be sold that does not conform to this standard. This is an example of a standard being enforced as law in order to protect the public.

You will look at several standards in the course of this block. As you see how standards are constructed and applied in practice, you should be able to develop some answers to the question 'Why do we have standards?'

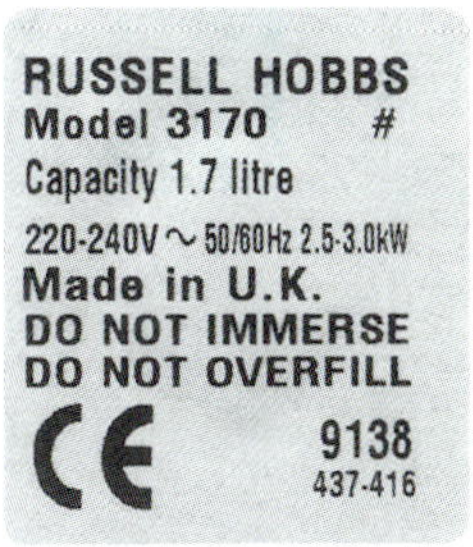

RUSSELL HOBBS
Model 3170   #
Capacity 1.7 litre
220-240V ~ 50/60Hz 2.5-3.0kW
Made in U.K.
DO NOT IMMERSE
DO NOT OVERFILL
CE   9138
          437-416

Figure 1.2 CE mark on a kettle

2.2 Developing standards

In the UK, standards are issued by the BSI (British Standards Institution). There are other national and international standards organizations, such as ASTM (American Society for Testing Materials), DIN (Deutsches Institut für Normung), ISO (International Organization for Standardization), and many others. These are not bodies which develop standards in isolation and impose them on a particular engineering community. The development of a standard is often driven by a group of people, working in a certain area, who want to produce a blueprint for a particular method which can be used by themselves and by their colleagues. Furthermore, standards are not set in stone once issued, but may be revised and updated. Supplements or new related standards may be produced if an existing standard is found to be insufficient to cover all aspects of a particular area, or if there is a change in practice in the field for which the standard is applicable.

In recent years, there has even been a drive towards standardization of the standards themselves. When different countries had different standards relating to the same product, with different criteria and different test methods, it placed a burden on manufacturers to indicate that their product complied with each individual standard if the product was sold in many countries. Many standards now have EN or ISO prefixes, to indicate that they have European or international applicability.

For example, in the next section we will look at a standard on eye protection. In the UK this standard existed formerly as BS 2092, but was revised in 1995 and renumbered as BS EN 166. Prior to this, there had been a plethora of standards within Europe covering this type of product.

The best way to learn about the purpose of a standard is to examine one. The standards I have chosen to look at contain information which is relevant to other sections of the course. You may, for the sake of interest, want to explore the wide range of available standards for yourself. You can do this in any large library, or simply by browsing the titles available from BSI's catalogue (available on-line at www.bsi-global.com). In general, because of the high cost of formulating a standard, both in terms of professional expertise and administrative effort, it is necessary to pay in order to obtain a complete copy of a standard.

2.3 Looking at a standard: eye protectors

The standards discussed in the following sections can be found on the CD-ROM.

In many workplaces involved directly with manufacturing, there is the possibility of hazardous debris being flung around from various sources. Obvious examples are when components are being cut or otherwise machined, or if hot or toxic liquids are being handled.

There is a wide range of safety equipment used in such environments. One of the most basic is eye protection – that is, goggles or a visor to prevent any hazard from damaging the eyes of a worker (Figure 1.3).

Eye protection equipment is sufficiently important to have a standard dedicated to it (designated BS EN 166: recall that this means that the standard is a British Standard, and that it also applicable across Europe). This means that anyone purchasing protective eyewear, whether for a school chemistry laboratory or a production workshop, can check that the product conforms to this standard in order to know that it will indeed be suitable for the job for which it is intended. The product will bear a label indicating its suitability.

BS EN 166 is reproduced on the CD-ROM

If we look at what BS EN 166 covers, we find it says the following.

0 Introduction

This standard deals with general considerations relating to eye-protectors, such as:

- designation;

- classification;

- basic requirements applicable to all eye-protectors;

- various particular and optional requirements;

- allocation of requirements, testing and application;

- marking;

- information for users.

1 Object

This standard specifies functional requirements for various types of personal eye-protectors.

The transmittance requirements for various types of filter oculars are given in separate standards (see clause 3).

2 Scope

This standard applies to all types of personal eye-protectors used against various hazards, as encountered in industry, laboratories, educational establishments, DIY activities, etc. which are likely to damage the eye or impair vision, with the exception of nuclear radiation, X-rays, laser beams and low temperature infra-red (IR) radiation emitted by low temperature sources.

The requirements of this standard do not apply to eye-protectors for which separate and complete standards exist, such as laser eye-protectors, sunglasses for general use, etc. unless such standards make specific reference to this standard.

Eye-protectors fitted with prescription lenses are not excluded from the field of application. The refractive power tolerances and other special characteristics dependent upon the prescription requirement are specified in ISO/DIS 8980-1 and ISO/DIS 8980-2.

(You may want to have a quick look at the entire standard on the CD-ROM. Don't try to read through the whole document, as it contains a lot of detail that you don't need to be aware of. For now, just look at the headings of what it contains: but see ▼**Writing decimal numbers**▲ and ▼**Precision**▲.)

SAQ 1.1 (Learning outcome 1.1)

Is BS EN 166 applicable for the following products?

1 A set of goggles worn by a welder.

2 Protective glasses worn by a worker at a nuclear reprocessing plant.

3 Sunglasses for use in cold weather.

4 Prescription glasses (i.e. glasses to correct for a vision problem like short-sightedness) supplied to a worker who operates a lathe.

SAQ 1.2 (Learning outcome 1.1)

Is BS EN 166 the only standard in this field?

Figure 1.3 Reconstruction of an accident in which polycarbonate lenses prevented serious damage to the wearer's eye

▼Writing decimal numbers▲

In the English-speaking world, fractional decimal numbers are conventionally written with a so-called 'decimal point' after the whole number. So $2\frac{1}{2}$ is written

as 2.5. (See the *Sciences Good Study Guide* Maths Help p. 349 for more on decimal numbers.) This is the notation which we have already used as the norm for this course.

In some European countries, the decimal point is often replaced by a comma; so in this example 2.5 would be written as 2,5. As BS EN 166 is a pan-European standard, decimal numbers where there is a figure after the decimal point are given using the comma notation.

▼Precision▲

It is not unusual in standards, or any other sort of engineering specification, to see a length quoted as something like: length = 20 mm ± 0.5 mm. The ± symbol is a combination of a plus and a minus sign, and is read as 'plus or minus'.

A specified length of 20 mm ± 0.5 mm (say) means that any length from 19.5 mm to 20.5 mm will meet the specification. The 'plus or minus' part of the specified length is sometimes called the tolerance. The part of the specification that comes before the tolerance is the nominal value. In this case the nominal value is 20 mm.

Any dimensions, or other measurements, in a specification are likely to be given as a nominal value and a tolerance. There are at least two reasons for this.

1   In practice, an exact nominal value is usually not required. For instance, a range of sizes for a particular component may give perfectly satisfactory operation when the component is used in a machine. It is more useful to the manufacturer of the component to say what this range is than to stipulate an exact length without a tolerance.

2   There would be no point stipulating an exact length anyway, since there is always a degree of uncertainty in any measurement. A manufacturer could never be certain that a component met the required specification if all that had been specified was a nominal value.

A tolerance of ± 0.5 mm on a nominal length of 20 mm would not be considered unusual in many undemanding applications. A tolerance of ± 2 mm on the same nominal length would be quite wide, as 2 mm is 10% of 20 mm.

Another way of expressing a length of 20 mm ± 0.5 mm would be (20 ± 0.5) mm. Sometimes the tolerance is not spread evenly on either side of the nominal value. For example, a particular speed might be given as:

$$12.0_0^{+0.6} \text{ m s}^{-1}$$

This means that the speed can be faster than the nominal speed of 12 m s^{-1} by up to 0.6 m s^{-1}, but *may not be less*. This particular notation is fairly unusual, and you will not encounter it again in the course.

The BS EN 166 standard is quite broad in its coverage of the requirement for a set of eye protectors. Eye protectors might have to be exposed to flying objects, small particles, splashes of liquid metal, and flashes of light bright enough to cause blindness. All of these hazards are encompassed by the standard, although it is not necessary for a particular product to be safe against every single one of them.

I will look at the criterion which is set for robustness, which is tested as follows:

> The requirement for minimum robustness is satisfied if the ocular withstands the application of a 22 mm nominal diameter steel ball with a force of (100 ± 2) N.

The ocular is the piece of material which forms the lens of the protective eyewear; that is, the clear barrier in front of the eye.

The above extract states the *minimum* robustness. There is also an enhanced robustness criterion. If the ocular is removed from the spectacle frame for testing, this criterion is:

> The oculars shall withstand the impact of a 22 mm nominal diameter steel ball, of 43 g minimum mass, striking the ocular at a speed of approximately 5.1 m/s.

In both cases, the ball must not penetrate the lens, and the lens must not fracture or be greatly deformed if the protectors are to pass the test. If the ball passes through the lens or the lens breaks, then clearly the protectors would not be doing their job. If the lenses are tested whilst mounted in their frames, then they must not be pushed out of the frame, even if they remain intact when doing so.

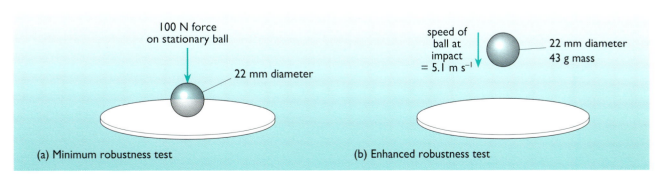

(a) Minimum robustness test

(b) Enhanced robustness test

Figure 1.4 Illustrations of the tests

Exercise 1.1 (revision)

The volume V of a spherical object (a ball) is given by

$$V = \frac{4}{3}\pi r^3$$

where r is the radius (which is half of the diameter) of the sphere. (See ▼π; a fundamental constant▲.) The density of steel is 7800 kg m^{-3}. Use this information to calculate the mass of the 22 mm diameter steel ball described above.

More information on volumes is provided in the *Sciences Good Study Guide* Maths Help, from p. 392.

SAQ 1.3 (Learning outcome 1.2)

Imagine you are going to test a spectacle lens against BS EN 166 for

1 minimum robustness,

2 enhanced robustness.

What quantities must you be able to measure accurately for each of these two sets of tests?

From the introduction to this standard, and the answer to SAQ 1.3, we can see that whilst any individual standard may appear self-contained, there is often subsidiary information which is needed by a manufacturer. BS EN 166 does not indicate *how* the measurements are to be performed, only what the requirements are. If the protective glasses are fitted with prescription lenses, there are different standards which also need to be taken into account.

In fact, the specifications for robustness given above specify that the test methods should be in accordance with the method given in another standard, BS EN 168. So, in addition to a standard on the requirements for protective lenses, there is also a standard which gives information on how these requirements may be ensured. One of the reasons for having standards is that they provide a way of disseminating advice and information on how to complete basic measurements, or use a particular product in a certain application.

Towards the beginning of BS EN 166, there is a section (Section 3) which indicates what standards are related to it: these are shown below.

3 Normative references

This European Standard incorporates, by dated or undated reference, provisions from other publications. These normative references are cited at the appropriate places in the text and the publications are listed hereafter. For dated references, subsequent amendments to or revisions of any of these publications apply to this European Standard only when incorporated in it by amendment or revision. For undated references the latest edition of the publication referred to applies.

EN 165: Personal eye-protection – Vocabulary

EN 167 : 1995 Personal eye-protection – Optical test methods

EN 168 : 1995 Personal eye-protection – Non-optical test methods

EN 169 : Personal eye-protection – Filters for welding and related techniques – Transmittance requirements and recommended utilisation

EN 170 : Personal eye-protection – Ultraviolet filters – Transmittance requirements and recommended use

EN 171 : Personal eye-protection – Infra-red filters – Transmittance requirements and recommended use

EN 172 : Personal eye-protection – Sunglare filters for industrial use

EN 379 : Specification for welding filters with switchable luminous transmittance and welding filters with dual luminous transmittance

ISO/DIS 8980-1: Ophthalmic optics – Uncut finished spectacle lenses – Part 1 : Specifications for single vision and multifocal lenses

ISO/DIS 8980-2 : Ophthalmic optics – Uncut finished spectacle lenses – Part 2 : Specifications for progressive power spectacle lenses

Some of the related standards contain test methods (such as *Personal eye protection – Non-optical test methods*); others cover eye protection for more specific use, such as for sunglare filters. It soon becomes clear that no standard exists in isolation. The list of standards relating to eye protection is a good example. The standards for personal eye protection differ from those for spectacle lenses, and differ from those relating to protection from strong light sources – whether this be the sun or an artificial source of radiation. (See ▼**Radiation, light and the visible spectrum**▲.)

▼π; a fundamental constant▲

When dealing with circles and spheres, you will often see the symbol π cropping up. This is the Greek letter pi (pronounced *pie*). It is used to represent the ratio between the circumference of a circle and its diameter. (The circumference of a circle is the distance you travel in one complete circuit of the edge of the circle.) The ratio of circumference/diameter is the same for *every* circle, and is about 3.14. This ratio is called π, and its exact value is impossible to determine with complete precision because the decimal part of the number is infinitely long. It has

been calculated to some millions of decimal places, and still it keeps going. However, for our purposes, the value of 3.14 for π is adequate. Some more information on π is given in the *Sciences Good Study Guide* Maths Help, from p.386.

Exercise 1.2 (revision)

What are the units of π?

▼Radiation, light and the visible spectrum▲

Colloquially, the word *radiation* is taken to mean the harmful emissions from radioactive material. However, in its correct scientific usage, radiation refers to various types of emission, only some of which are related to radioactivity, and only some of which are harmful. One of the commonest types of radiation is *electromagnetic* radiation, which is itself an entire spectrum of kinds of radiation, including for example light, microwaves, X-rays and some forms of damaging radioactivity. All varieties of electromagnetic radiation have the same fundamental nature, which is what the term 'electromagnetic' refers to. (Electromagnetic radiation is distinguished from particle radiation, which has a different nature.) We can use the example of light to illustrate properties that are common to all electromagnetic radiation.

In simple terms, light can be envisaged as travelling as a wave, rather like the way the ripples in a pond travel out from their source. The distance between the peaks of the waves (Figure 1.5) is known as the *wave length*, or just wavelength. Wavelength can vary enormously, and as the wavelength changes so the properties of the radiation change. Since we classify various forms of electromagnetic radiation by their properties, the wavelength provides a useful way of specifying particular types of electromagnetic radiation. Figure 1.6 shows the electromagnetic spectrum. You can see that it ranges from gamma rays (one of the harmful types of radiation) at very short wavelength, through to radiowaves at long wavelengths. Visible light comes somewhere in between. Once again, we see the use of a *logarithmic* scale for plotting data: the wavelengths shown on this figure range from 10^{-13} m at the gamma-ray end (equivalent to a tenth of a thousandth of a billionth of a metre) to 10^6 m (one thousand kilometres) at the radiowave end.

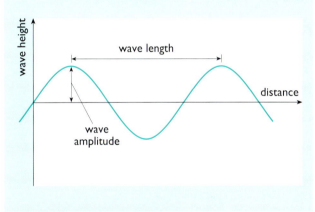

Figure 1.5 Waves and wavelength

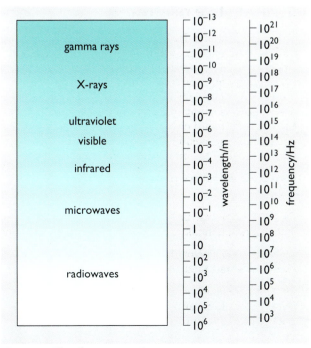

Figure 1.6 The electromagnetic spectrum

Another useful way of characterizing electromagnetic radiation (or any other wave phenomenon, such as sound) is by its frequency. This is the number of waves that pass a stationary point each second, measured by counting the peaks per second. Although frequency is a simple concept – a number per second – it is given its own unit, the hertz, abbreviated to Hz. A frequency of 10 Hz is ten complete waves per second.

Exercise 1.3

From Figure 1.6:

1   What is the approximate wavelength of visible light?

2   What is the frequency of microwaves?

Let's now investigate the detailed procedure for the increased robustness test. BS EN 166 specifies that the test ball must strike the ocular with a certain speed: around 5.1 m s^{-1} (Figure 1.4). The standard also indicates that dropping the ball from a height of 1.3 m will provide this speed.

Of course the standard makes the assumption that the test is carried out on Earth, and that you haven't done something bizarre like cover the test rig in treacle! From Block 1, you know that gravity produces a force on everything at the Earth's surface, and when something is dropped this force causes the object to accelerate. You may also remember that the acceleration from gravity is 9.8 m s^{-2}.

Exercise 1.4 (revision)

If an object is dropped on Earth from rest, how fast will it be travelling:

1   after 1 second?

2   after 5 seconds?

Exercise 1.4 showed that it was possible to calculate how fast an object would be travelling after it had been accelerating for a certain length of time. We can also calculate how fast an object would be travelling after it had been accelerating over a certain distance.

If we use some symbols to represent speeds, acceleration and distance, we can make this relationship easier to write. If there is an acceleration a over a distance s, then something that starts with a speed of u will end up with a speed of v. The mathematical derivation of the relationship between all these is outside the scope of this course, but the answer turns out to be:

$$v^2 = u^2 + 2as$$

Thus if we know u, the starting speed of the object at a moment in time, we can work out what the final speed v will be after an accelerating or decelerating force acts upon it over the distance s.

In the case of a ball which is dropped, there is zero initial speed, so our equation becomes:

$$v^2 = 2as$$

which can also be written as:

$$v = \sqrt{2as}$$

The term v is the final speed after the object has moved through the distance s (fallen this distance, in the case of an object being dropped). The term a represents is the acceleration that it experiences. Properly speaking, a represents the *magnitude* of the acceleration. You will recall that acceleration is a vector, and therefore incorporates both a magnitude and a direction. The above equation takes no account of the direction of the acceleration.

Information on using roots is given in the *Sciences Good Study Guide*, Maths Help, from p. 356.

SAQ 1.4 (Learning outcome 1.5)

Confirm that a ball dropped from a height of 1.3 m will be travelling with a speed of 5.1 m s^{-1} when it hits the surface below if air resistance is ignored.

Fortunately we have been able to use a bit of fundamental physics here – the fact that a ball dropped under Earth's gravity accelerates at a certain rate – to help us establish the speed of the ball. If we had had to measure the speed of the ball directly in order to prove that the test conditions were correct, it would have involved complicated and probably expensive equipment, all of which would have added to the final cost of the product. One of the key points that I want to convey here is that measurement methods are important. We have discussed already how the units of measurement are critical in ensuring compatibility of data. The way in which the measurements are made, in order to find values for specific quantities, is equally critical. This is why the eye protection standard has related standards concerned solely with the methods of test.

You may have noticed, if you have looked at the standard, that impact resistance is not the only protection offered by eye protectors. They can also give protection against dust, and splashes of liquid metal, for example. Each of these elements of protection is tested in a particular way.

The consequence of all this is that as long as the conditions of the test, which are specified by the standard, are adhered to, the results for a particular make of eye protector should be the same no matter where the test is conducted. The fact that there is a standard for the test method as well as for the specific criteria for the protectors indicates the importance of the testing method. Without standardization of this kind, it would be possible for manufacturers to claim that their product offered superior protection, whilst using non-rigorous testing methods to define their performance. As it is, eye protectors which conform to BS EN 166 will be labelled as having conformed to the relevant requirements.

SAQ 1.5 (Learning outcome 1.5)

A new apparatus for firing ball bearings horizontally has been developed for testing the lenses of eye protectors. The balls are accelerated at a rate of 80 m s^{-2} for 0.1 s, and then travel 2 m before hitting the lens. While they are travelling, there is a deceleration (a negative acceleration) of 6 m s^{-2} owing to air resistance.

What will be the horizontal velocity of the ball be when it reaches the lens? Is this consistent with the requirements of BS EN 166?

2.4 Summary of Section 2

This section has introduced you to the sorts of requirement which are aimed at in a standard for a particular *product*. The case we have examined, the standard for eye protectors, gives specific tests that the product must pass before it can claim to conform to the standard. This is typical of a standard for a product or a range of products. Such standards tend to define the function and performance of the product, and there may be tests which the manufacturers must apply, or certain ranges of dimensions like length or weight, which the product must comply with.

This is just one type of standard. Examples of other types of standards are those dedicated to testing methods, and standards for processes or practices. We shall look at examples of these in the subsequent sections.

3 A standard for a test method

In Block 1 you came across the concept of different materials having different properties. Flint, bronze and iron offered progressively better properties for our ancestors to exploit in the form of tools, but it is only relatively recently, with advances in alloy technology and the advent of polymers, that we have had a plethora of materials at our disposal. In Block 2, we looked at how different materials can be compared to try to select the best material for a certain application. In that case, the property that was important was the elastic modulus (Young's modulus), the stiffness, of the material. For properties such as this, I hope you will appreciate that there is a requirement for standardization in their measurement, or testing, in the same way that the robustness of eye protectors requires a standard.

A designer working on a product will often have to specify more than one material property. Which properties are important will depend on the particular application. For a lift cable it is fairly clear that strength will be critical, but there are often other properties that need to be considered, depending on the application that the material is required for. So the materials used in a computer microprocessor chip are carefully selected and tailored for specific electrical and electronic properties; and a turbine blade for an aeroplane engine must have good mechanical properties at temperatures exceeding 1000 °C. And so on.

SAQ 1.6 (Learning outcome 1.5)

What sort of properties do you think are important for the *materials* selected for the following applications? You may be able to come up with more than one answer to each.

1   The surface of a motorway.
2   The core of an electrical mains cable.
3   The insulation around an electrical mains cable.
4   The wing of an aeroplane.

In this section, I'm going to look at the standard for measuring the strength of a particular class of materials, namely metals. You will see that, as with the eye protectors, there is a set of carefully prescribed conditions for undertaking such a test.

The standard in question is BS EN 10002 *Tensile testing of metallic materials.* Once again, the standard is applicable across Europe. In this case, the standard is split into several parts, in a similar way to the eye-protection standard, with reference out to specifics for certain test methods. The parts are:

- BS EN 10002-1:1990: *Tensile testing of metallic materials. Method of test at ambient temperature*
- BS EN 10002-2:1992: *Tensile testing of metallic materials. Verification of the force measuring system of the tensile testing machine.* This particular standard has actually since been superseded by BS EN ISO 7500-1:1999: *Metallic materials. Verification of static uniaxial testing machines. Tension/compression testing machines. Verification and calibration of the force-measuring system.* The revised standard has international applicability.
- BS EN 10002-3:1995: *Tensile testing of metallic materials. Calibration of force proving instruments used for the verification of uniaxial testing machines.*
- BS EN 10002-4:1995: *Tensile testing of metallic materials. Verification of extensometers used in uniaxial testing.*
- BS EN 10002-5:1992: *Tensile testing of metallic materials. Method of test at elevated temperatures.*

Two of the parts only (parts 1 and 5) deal with the actual business of the testing. One is for ordinary room temperature (or 'ambient' temperature) testing, and the other deals with testing at higher temperatures. The other sections of the standard are to do with checking that the various parts of the testing equipment are giving the right readings. This is equivalent to ensuring that your metre rule or kilogram weight are actually in accord with the defined standard.

The full text of the standard is on the CD-ROM.

Let's look at the coverage of BS EN 10002-1.

1 Object and field of application

This European Standard specifies the method for tensile testing of metallic materials and defines the mechanical properties which can be determined thereby at ambient temperature.

For certain particular metallic materials and applications, the tensile test may be the subject of specific standards or particular requirements.

3 Principle

The test involves straining a test piece by tensile force, generally to fracture, for the purpose of determining one or more of the mechanical properties defined in clause 4.

The test is carried out at ambient temperature between 10 °C and 35 °C, unless otherwise specified. Tests carried out under controlled conditions shall be made at a temperature of 23 ± 5 °C.

SAQ 1.7 (Learning outcome 1.1 and revision)

1. The standard specifies that it is applicable to metallic materials. What classes of material does this exclude?
2. The standard refers to tensile testing, i.e. the sample is subjected to a tension force. What other type of force does this exclude?

The standard covers tensile loading only because the measurement of properties under compression loading makes for a more complicated test. Samples can buckle under compressive loading, and this can happen before the apparent 'real' strength limit of the material is reached.

Compression testing and bending testing are, however, often more appropriate for ceramic materials, whose strength is better in compression, and which can fail in the gripped region during a tension test because of local damage introduced by abrasion between the ceramic sample and the grips themselves.

A typical design for a specimen to be tested in tension is shown in Figure 1.7. The ends of the sample have to be compatible with the fittings of the testing machine to be used, allowing a force to be applied up to failure of the material without having the sample slip from the grips. Ideally, the centre region of the sample should have a smaller diameter than the grip ends, and the shape of transition between these regions is given in annexes to the standard.

Figure 1.7 Sample for tensile testing

Exercise 1.5 (revision)

Why is the middle of the sample a smaller diameter than the gripped ends? (*Hint*: think about the variation of stress in the sample.)

The test method is then fairly straightforward. An increasing force is applied to the sample, by pulling the grips apart, until the sample fails. The force is measured. The extension of the sample is measured also.

Typically, for a metal such as aluminium, the results from the test will look like the plot in Figure 1.8, which is the kind of graph you have already seen in Block 2.

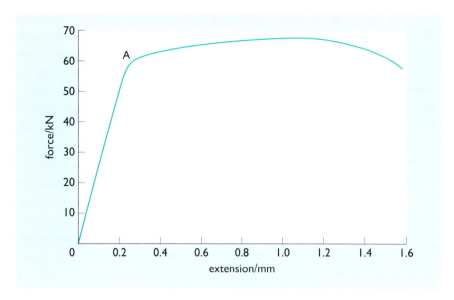

Figure 1.8 Typical force–extension curve for a metal

We use a graph to show the data because this allows much easier interpretation than if it were presented as numbers in a table of figures. By plotting a graph we can instantly see important features of how the sample behaves when the force is applied.

Unlike the similar graphs in Block 2, this one shows what happens if the sample is taken all the way to failure. The graph splits into two sections. Up to the point marked 'A' the graph is linear: a straight line (this is the bit we looked at in Block 2). Up to this point, an increase in force produces a proportional increase in extension. Beyond this point, a small increase in load produces a large increase in extension: the metal is exhibiting ▼**Plasticity**▲. Finally, the load begins to reduce as the sample extends to the final failure point.

The type of graph shown in Figure 1.8 would only be useful to make comparisons between different materials if we always used samples of exactly the same dimensions. Clearly this isn't practical, so we have to put some effort into analysing the data to get at the fundamental materials properties.

Exercise 1.6 (revision)

You should recall from Block 2 that force and extension are affected by the particular geometry (or shape) of the specimen being tested. What information is required in order to make conversions to stress and strain?

By looking at the shape of our test sample (look again at Figure 1.7), we can see how to get the information that is needed. The central part of the sample is known as the *gauge* or *gauge length*. The cross-sectional area of this part of

▼Plasticity▲

The word 'plastic', as we noted in Block 1, is generally associated with polymer products: plastic bags, plastic kettles, plastic toys. When referring to the mechanical behaviour of materials, the word has a completely different meaning. As shown in Figure 1.8, a typical metal will pass a point where a given increase in applied load causes a much bigger increase in extension than when the load was lower. What is more, the extra deformation produced is *permanent*. If you flex a

paperclip, you will find that after a small force is applied it will spring back to shape; but you can supply sufficient force to bend it completely: the metal has become plastic, and you have introduced a permanent deformation.

Plasticity is a feature of many polymer materials – hence the term 'plastics'. Ceramic materials do not show plasticity under normal conditions. They will fail by fracture without any plastic deformation.

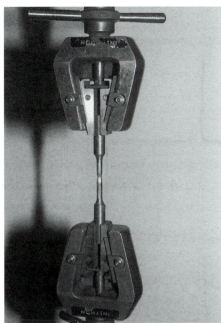

Figure 1.9 Metal tensile sample before testing (left) and once necking has occurred (right)

the sample is obtained easily from a measurement of its diameter. (Remember that the stress is highest in the gauge length of the sample, so this is where the sample should fail.) The length of the gauge is also easily measured: note that it is only the length of the gauge – the centre section – which is used to calculate the strain, as it is only over this region that the relevant extension occurs.

One of the consequences of the plastic behaviour described earlier is that a metal sample will generally form a 'neck' before failing. This is shown in Figure 1.9. The sample shows a thinning, or a reduction in cross-sectional area, at the point where ultimately it will fail. You can demonstrate this effect for yourself by pulling on the polymer binding used for a four-pack of drink cans (Figure 1.10). Eventually it will start to deform plastically, which means that it would not return to its original shape and length if you stopped pulling. With continued pulling, a region develops where the width of material narrows considerably, which is known as *necking*.

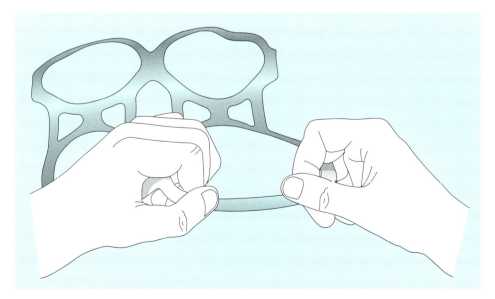

Figure 1.10 Pulling the polymer binding from a pack of drinks cans

The onset of necking explains why there is a reduction in load towards the end of the test. Because the neck has a smaller cross-section than the rest of the sample, the stress within the neck is higher than elsewhere, in the same way that the stress in the gauge is lower than the stress in the grips. Because the stress is higher, this gives more strain. Also, the force required to maintain the maximum stress is less, and this causes the drop in force which is measured. (Remember the test machine does not 'apply a force' as such: it generates a force in the sample by pulling the grips apart.)

So, using the calculations outlined earlier, Figure 1.8 can be converted to a stress–strain plot (Figure 1.11).

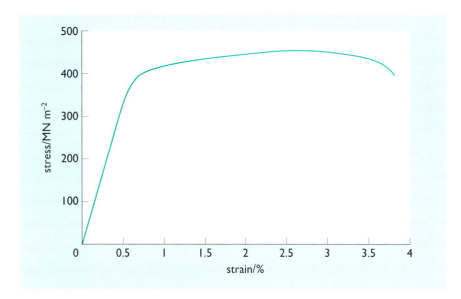

Figure 1.11 Stress–strain plot for a metal

Note that this plot also shows a reduction (in stress) towards the end of the test. The conversion to stress from force is undertaken using the initial cross-sectional area, and no account is taken of what happens during necking. Such a stress is known as the *engineering stress*, and it doesn't affect the use of the graph, as no stress-relevant information is obtained from that portion of the plot where the force begins to reduce following the start of necking.

The standard explains how such a plot is analysed to allow measurement of, for example, the tensile strength of the material. (The 'tensile strength' is the stress corresponding to the highest point on the stress–strain curve.) It is also possible to characterize a material by its *yield strength*. There are several ways that this can be approached, all intended to identify a reproducible way of quoting the stress which causes the material to become plastic.

One way is to calculate a so-called *proof stress*. The exact point at which the sample becomes plastic is very difficult to determine. Far easier is to identify a point where a certain amount of permanent deformation has occurred, and to quote this as a reference. An example of this is shown in Figure 1.12.

I have drawn a graph line passing through the origin of the graph (where stress and strain are zero). The slope of this line tells us the Young's modulus of the material, as you will recall from Block 2. If the material only ever showed elastic behaviour (no plasticity), there would not be any deviation from this line.

Suppose the material were loaded enough to take it into its plastic region, but not enough to cause the material to fail. Now when the load is removed the material will return to zero stress, but will have a permanent plastic deformation, or strain, depending on how far into the plastic region the material was taken. That is to say, for a sample permanently deformed in this way, zero stress would no longer correspond to zero strain. Rather, it would correspond to a certain amount of permanent strain, such as 0.1% strain or 0.15%, depending on how much the material had been deformed.

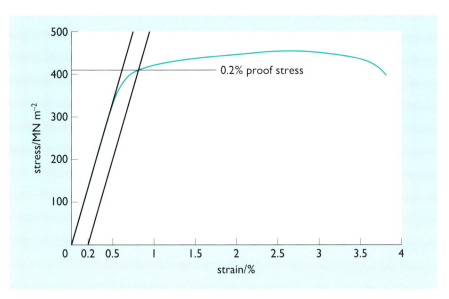

Figure 1.12 The data from Figure 1.11, showing calculation of the 0.2% proof stress

For a proof-stress calculation, we select a certain amount of plastic deformation (0.2% in this case) and find the stress required to cause it. At the 0.2% strain point on the strain axis (Figure 1.12), I construct a line that is parallel to the elastic region of the graph of the undeformed material. Where this second line intersects the first graph line is the point corresponding to the 0.2% proof stress. The value of stress on the stress axis corresponding to this point is the 0.2% proof stress.

SAQ 1.8 (Learning outcome 1.6)

From Figures 1.11 and 1.12, find the following properties for the sample that was tested:

1   The tensile strength.

2   The 0.2% proof stress.

You might be surprised by the lack of prescription in the sections of the standard concerning how the test is conducted. For the eye protectors, a product sold for a profit, the tests are specifically defined and carefully described. For strength measurement, reasonably free rein is given to the type of testing machine used, the rate at which the sample is tested, and the methods used to monitor force and extension. The standard concerns itself far more with specifying the analysis procedures for the test data once it has been collected.

There are two main reasons behind this. First, engineering materials come in a range of forms: bar, rod, plate, billet etc. Specifying standard dimensions for all samples would not be particularly helpful unless the sample size was made particularly small. This could then introduce errors into the answer obtained. For example a 0.02 mm inaccuracy in machining a sample of 1 mm diameter is much greater, as a percentage, than the same inaccuracy in a sample of 10 mm diameter.

Secondly, what is being tested is a *material property*. As long as care is taken over the measurement of the force (which gives the stress) and the extension (which gives the strain), then the answer should be correct regardless of the sample geometry and the precise specifications of the test machine. So in this case the standard is concerned more with ensuring that the data are analysed correctly than with precise definition of how the test is conducted.

4 Standards for processes

So far we have looked at two different types of standard: a standard for a product and a standard for a type of measurement. Another important class of standards are those that cover processes; by this I mean any series of steps that leads to the creation of a product.

You have already encountered an example of this in Block 2, where you saw a suggested model of the design process from BS 7000: *Design management systems*. Although that standard outlines a method for 'good' design, it does not directly show a designer how to create a new idea or to progress from a concept to a successful product.

In this section I will look at a similar type of standard, one associated with quality. Quality assurance has become an important theme throughout business, as a way of encouraging companies to review how they achieve a desired product for their customers. There is a swathe of standards dealing with quality systems and related issues.

The standard we shall look at to illustrate this type is BS EN ISO 9000-1: *Quality management and quality assurance standards. Guidelines for selection and use*. I think that it is informative in the first instance to look at Section 4 of this standard, 'Principal concepts', and I have extracted some paragraphs from this below. Read through this extract, and then consider the following self-assessment question. (The full standard is reproduced on the CD-ROM.)

4 Principal concepts

4.1 Key objectives and responsibilities for quality

An organization should:

(a) achieve, maintain and seek to improve continuously the quality of its products in relationship to the requirements for quality;

(b) improve the quality of its own operations, so as to meet continually all customers' and other stakeholders' stated and implied needs;

(c) provide confidence to its internal management and other employees that the requirements for quality are being fulfilled and maintained, and that quality improvement is taking place;

(d) provide confidence to the customers and other stakeholders that the requirements for quality are being, or will be, achieved in the delivered product;

(e) provide confidence that quality system requirements are fulfilled.

4.3 Distinguishing between quality system requirements and product requirements

The ISO 9000 family of International Standards makes a distinction between quality system requirements and product requirements. By means of this distinction, the ISO 9000 family applies to organizations providing products of all generic product categories, and to all product quality characteristics. The quality system requirements are complementary to the technical requirements of the product. The applicable technical specifications of the product (e.g. as set out in product standards) and technical specifications of the process

are separate and distinct from the applicable ISO 9000 family requirements or guidance.

International Standards in the ISO 9000 family, both guidance and requirements, are written in terms of the quality system objectives to be satisfied. These International Standards do not prescribe how to achieve the objectives but leave that choice to the management of the organisation.

4.5 Facets of quality

Four facets that are key contributions to product quality may be identified as follows:

(a) Quality due to definition of needs for the product

The first facet is quality due to defining and updating the product, to meet marketplace requirements and opportunities.

(b) Quality due to product design

The second facet is quality due to designing into the product the characteristics that enable it to meet marketplace requirements and opportunities, and to provide value to customers and other stakeholders. More precisely, quality due to product design is the product design features that influence the intended performance within a given grade, plus product design features that influence the robustness of product performance under variable conditions of production and use.

(c) Quality due to conformance to product design

The third facet is quality due to maintaining day-to-day consistency in conforming to product design and in providing the designed characteristics and values for customers and other stakeholders.

(d) Quality due to product support

The fourth facet is quality due to furnishing support throughout the product life cycle, as needed, to provide the designed characteristics and values for customers and other stakeholders.

For some products, the important quality characteristics include dependability characteristics. Dependability (i.e. reliability, maintainability and availability) may be influenced by all four facets of product quality.

A goal of the guidance and requirements of the International Standards in the ISO 9000 family is to meet the needs for all four facets of product quality. Some facets of quality may be specifically important, for example, in contractual situations but, in general, all facets contribute to the quality of the product. The ISO 9000 family explicitly provides generic quality management guidance and external quality assurance requirements on facets (a), (b), (c) and (d).

SAQ 1.9 (Learning outcomes 1.1 and 1.3)

1   Is this standard applicable to a particular family of products or processes?

2   Is the standard more applicable to a customer or to a supplier?

The 'principal concepts' indicate that the standard is completely broad in its applicability: there is no exclusion, implicit or explicit, of any area of business. The function of the standard is to disseminate 'best practice', so that customer and supplier have confidence in the quality of the specified product. The rationale for a standard like this is simple: if you can prove that you have followed the recommendations of such a standard, a potential customer will have faith in your operation, no matter what it produces.

You should note that the standard recognizes the importance of product design. A 'poor' design, which does not meet its functional specification, or which does meet the specification but which will fail in some aspect such as ease of use, means that the product is of inherently poor quality.

The related standards, such as ISO 9001, 9002 and 9003, are more specific in the methods by which quality assurance can be undertaken. The CD contains also a copy of BS EN ISO 9003: *Quality systems. Model for quality assurance in final inspection and test.* I have again extracted some sections from this below, which you can read and then consider the following self-assessment question.

1 Scope

This International Standard specifies quality system requirements for use where a supplier's capability to detect and control the disposition of any product nonconformity during final inspection and test needs to be demonstrated.

It is applicable in situations when the conformance of product to specified requirements can be shown with adequate confidence providing that certain suppliers' capabilities for inspection and tests conducted on finished product can be satisfactorily demonstrated.

4 Quality system requirements

4.1 Management responsibility

4.1.1 Quality policy

The supplier's management with executive responsibility shall define and document its policy for quality, including objectives for quality and its commitment to quality. The quality policy shall be relevant to the supplier's organizational goals and the expectations and needs of its customers. The supplier shall ensure that this policy is understood, implemented and maintained at all levels in the organization.

4.10.1 General

The supplier shall establish and maintain documented procedures for final inspection and testing activities in order to verify that the specified requirements for finished product are met. The required final inspection and testing, and the records to be established, shall be detailed in the quality plan or documented procedures.

4.11 Control of inspection, measuring and test equipment

4.11.1 General

The supplier shall establish and maintain documented procedures to control, calibrate and maintain final inspection, measuring and test equipment (including test software) used by the supplier to demonstrate the conformance of product to the specified requirements.

Inspection, measuring and test equipment shall be used in a manner which ensures that measurement uncertainty is known and is consistent with the required measurement capability.

4.11.2 Control procedure

The supplier shall:

(a) determine the measurements to be made and the accuracy required, and select the appropriate inspection, measuring and test equipment that is capable of the necessary accuracy and precision;

(b) identify all inspection, measuring and test equipment that can affect product quality, and calibrate and adjust them at prescribed intervals, or prior to use, against certified equipment having a known valid relationship to internationally or nationally recognized standards. Where no such standards exist, the basis used for calibration shall be documented;

(c) define the process employed for the calibration of inspection, measuring and test equipment, including details of equipment type, unique identification, location, frequency of checks, check method, acceptance criteria and the action to be taken when results are unsatisfactory;

(d) identify inspection, measuring and test equipment with a suitable indicator or approved identification record to show the calibration status;

(e) maintain calibration records for inspection, measuring and test equipment (see 4.16);

(f) assess and document the validity of previous inspection and test results when inspection, measuring or test equipment is found to be out of calibration;

(g) ensure that the environmental conditions are suitable for the calibrations, inspections, measurements and tests being carried out;

(h) ensure that the handling, preservation and storage of inspection, measuring and test equipment is such that the accuracy and fitness for use are maintained

(i) safeguard inspection, measuring and test facilities, including both test hardware and test software, from adjustments which would invalidate the calibration setting.

SAQ 1.10 (Learning outcomes 1.1 and 1.3)

1   Does the standard consider only final testing of products? What is its scope?

2   Section 4.11.2 of ISO 9003 covers the 'control procedure' for quality assurance of a product. How does this section pertain to the standard for eye protectors?

Quality standards are a type of standard which are very broad in their application. This is completely different from the product-based safety standards which we looked at first, and the test-method based standards. In each case the function of the standard is different, although an engineer working in a particular sector would likely have a familiarity with each general type.

This brings us back to where I started at the beginning of this block: uniformity of product. The uniform dimensions required for interchangeable parts for musket locks is not something that was ever covered by a standard; nor are most of the products which are sold between companies and on to consumers specified by standards. However, the application of a quality standard to the production process helps to ensure reproducibility, and can give guidance on the systems that can be implemented to aid this process.

For all the standards that I've covered, the most important function they have is communication: communication of how to perform a test, or how to check the safety of a product, or how to go about ensuring that your customers are content with the quality of the product or service that they receive.

5 Other types of standard

Of course, not all standards are official documents that have BS or ISO numbers. Agreements on codes of practice, which are used to ensure the quality of products in particular industries, are often in widespread use without having the stamp of being a published 'standard'. There is not necessarily any differentiation of function between such documents and 'standards' proper; rather it is just that the companies involved have never sought to develop a full standard, and there has been no drive to do so from any other source.

Such codes of practice do not have any less engineering validity, however, and a manufacturer who chooses to ignore the wisdom contained in a code may be running risks of being sued for consequential damage (see ▼Failure of a polymer storage tank▲).

The existence of a standard or a code of practice implies that best practice in a certain area is known and documented. Hence ignorance of or non-compliance with such information could be regarded as negligent if a product failure leads to injury or loss.

▼Failure of a polymer storage tank▲

Figure 1.13 shows a polypropylene storage tank which was used to store concentrated caustic soda solution. This is used in the manufacture of various products such as soap and other detergents. In its concentrated state, however, it is a very unpleasant chemical, and can cause severe burns. The tank was constructed by welding together rings of polypropylene. It split at one of the welds.

![Figure 1.13 A polypropylene storage tank]

Figure 1.13 A polypropylene storage tank

There is no UK or European standard that directly covers the production of such tanks, but there is a German code of practice (designated DVS 2205). The tank that failed was intended to comply with the recommendations contained within this code, but didn't.

An analysis of the wall thickness of the tank was undertaken. According to the code of practice, the thickness of plastic should be greater at the bottom of the tank than at the top (simply because of the higher pressure at the bottom from the liquid contained in the tank). The code also contains calculations to show what the minimum wall thickness should be for tanks of a given diameter and height.

Figure 1.14 shows the wall-thickness profile of the real tank, along with the minimum wall thickness suggested by the code of practice. The green shaded area shows the regions where the thickness of the tank wall is less than that suggested by the code of practice.

It is easily seen that, at the point where the crack started, the wall thickness was less than half of what it should have been. Additionally, the presence of a weld tends to cause some weakening of a material, so it is particularly important that the wall thickness be sufficiently large to counter any deficiency at that point.

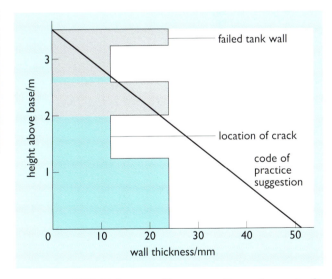

Figure 1.14 Wall thickness profile compared to requirements of the code

The code contained a designation of the minimum wall thickness for a good reason: to ensure that the tank would not fail. The use of thinner material was the root cause of the failure in this case.

6 Summary

Standardization is one of the underpinnings of a national (and, increasingly, global) engineering economy. It can come in a variety of forms. Some of these were introduced earlier in the course, like the SI units used for measurement of length, mass, force and so on. Standards of measurement are important for mass production, and in particular for modern mass production where parts for a product may be manufactured at different plants in different parts of the world.

At a higher level, standards are a means by which engineers can communicate best practice. For example, suppose you want to know how to measure the strength of a material. A good course of action is to read the appropriate standard, and then check that the equipment you are using has been calibrated to the appropriate standard itself.

Codes of practice may be developed jointly by engineers from different companies, who work in the same field. They are a way of sharing information by which a sector of industry can, overall, increase its knowledge base, and thereby hopefully increase efficiency or reduce the number of defective products.

Best practice does not have to be specific to a particular product or method, though. The modern drive for quality assessment and assurance has led to the production of standards which are generic in their applicability to virtually any customer/supplier relationship.

Finally, standards can be used as a protection to the consumer, by ensuring that products marketed with a particular function are fit for purpose. In this case, advertisement that the product conforms to a British or International Standard can act as a stamp of quality. However, whether or not a particular product is covered by a standard, a manufacturer can be fully liable for any injury caused by a fault. Consumers are protected by laws which go far beyond the enforceability of any standards.

Standards which have legal standing tend to be those associated with safety, and compliance with them is often mandatory under national legislation: safety of electrical equipment being a good example. Failure to comply with a standard can form the basis of legal action, if a manufacturer can be shown to have produced a poor or deficient product as a result. Alternatively, the provisions of a certain standard might be used as the basis of a contract between a supplier and a customer.

This is by no means a comprehensive description of the field of standards. You will quite possibly have experience of standardization which does not quite match any of the loose definitions that I have provided here. None the less, I hope that you can see the advantage of standards in the communication process which is essential for good engineering.

7 Learning outcomes

After completing your study of this section you should:

1.1 be able to interpret the applicability and scope of a standard;

1.2 be able to identify what is required to comply with a standard;

1.3 understand that different types of standard have differing functions, depending on whether they refer to products, properties, or processes;

1.4 be able to perform simple calculations relating to the requirements of a standard;

1.5 appreciate that materials have a wide range of useful properties, and standardization of methods for measuring these properties is required;

1.6 be able to interpret stress–strain plots to obtain information about the strength of a material.

Answers to exercises

Exercise 1.1

The volume of the ball is

$$\frac{4}{3}\pi r^3 = \frac{4}{3} \times 3.14 \times \left(\frac{0.022}{2}\right)^3 \mathrm{m}^3 = 5.57 \times 10^{-6}\,\mathrm{m}^3$$

We can now use this to calculate the mass of the ball.

$$\text{Density} = \frac{\text{mass}}{\text{volume}}$$

so

$$\text{Mass} = \text{density} \times \text{volume}$$

$$\text{Mass} = (7800 \times 5.57 \times 10^{-6})\,\mathrm{kg} = 0.0435\,\mathrm{kg} = 43.5\,\mathrm{g}.$$

So you can see that a steel ball, of the required diameter, *must* have a mass of at least 43 g.

Exercise 1.2

The ratio π is calculated by dividing circumference (a length) by diameter (another length). So these units cancel each other. Thus π has no units: it is just a number.

Exercise 1.3

1 Visible light has a wavelength of between 10^{-5} m and 10^{-6} m.

2 Microwaves have a frequency between 10^9 Hz and 10^{10} Hz.

Exercise 1.4

Under Earth's gravity, an object will accelerate at 9.8 metres per second per second.

1 After one second, it will be travelling at 9.8 m s^{-1}: It's as easy as that!

2 Every second, the object will speed up by a further 9.8 m s^{-1}, so after 5 seconds it will be travelling at (5×9.8) m s^{-1} = 49 m s^{-1}. The velocity is simply the acceleration multiplied by the time for which it acts.

Exercise 1.5

Where the diameter is smaller, the stress is larger. Hence the sample is most likely to fail in the centre section.

Exercise 1.6

To convert from force to stress, we must know the cross-sectional area of the sample.

To convert from extension to strain, it's necessary to know the original length of the sample.

Remember that notes and practice on algebra can be found in the *Sciences Good Study Guide* and on the CD-ROM 'Numeracy' section.

Answers to self-assessment questions

SAQ 1.1

1 Yes, this is within the scope of the standard.
2 No, as the standard does not cover nuclear radiation.
3 No, it is indicated that sunglasses are covered by a separate standard.
4 Yes, this is covered by the standard, in the final paragraph of the 'Scope'.

SAQ 1.2

No. The Scope indicates that there are separate standards covering, for example, eye protectors for laser light.

SAQ 1.3

1 For minimum robustness, it is necessary to know the ball's diameter, its mass, and the force used to press it into the ocular.
2 For enhanced robustness, rather than measuring force, it is necessary to measure speed, which, as we shall see, is actually done by measuring the height from which the ball is dropped.

SAQ 1.4

Using the equation

$$v = \sqrt{2as}$$

we find that:

$$v = \sqrt{2 \times 9.8 \times 1.3} \text{ m s}^{-1}$$

$$= \sqrt{25.5} \text{ m s}^{-1}$$

$$= 5.05 \text{ m s}^{-1}$$

This is just less than the 5.1 m s^{-1} expected.

SAQ 1.5

The ball is accelerated at 80 m s^{-2} for 0.1 s. The speed after this acceleration is just the acceleration multiplied by the time, so the speed will be:

$$80 \text{ m s}^{-2} \times 0.1 \text{ s} = 8 \text{ m s}^{-1}$$

The ball then slows down as it travels towards the lens. The final horizontal velocity can be derived from:

$$v^2 = u^2 + 2as$$

That is,

$$v = \sqrt{8^2 - (2 \times 6 \times 2)} \text{ m s}^{-1}$$

Note that the acceleration must be written as a negative number here, as the ball is being slowed by the air resistance.

$$v = \sqrt{(40)} \text{ m s}^{-1}$$

$$= 6.3 \text{ m s}^{-1}$$

This is above the speed recommended by the standard, so the test, although possibly more stringent than the requirements, would not conform. The machine could be adjusted to provide the correct speed.

SAQ 1.6

1   The surface of a motorway must have good wear resistance, so that the surface is long-lasting despite the passage of many vehicles.

2   An electrical cable must conduct electricity, so the core must be an electrical conductor.

3   The insulation for the cable clearly should not conduct electricity, to prevent electric shocks when the cable is being handled. The insulation should be fairly heat resistant, in case the cable is overloaded and begins to heat up.

4   An aeroplane wing requires strength, so that there is no danger of it failing under the loading experienced during flight. It also requires stiffness, so that it does not bend out of shape during flight to the extent where the aircraft would no longer fly.

In each case, there may have been other properties which were equally important. Think carefully about whether the property you chose is a property of the *material*, though, or whether it's just a consequence of the way the material is formed or processed.

SAQ 1.7

1   As the standard refers to metals, it is not appropriate for testing of ceramic or polymeric materials.

2   It excludes compressive force, which could be the basis of another type of test. (There are also complicated tests involving torsion, which is a twisting of the sample, that we won't cover in this course. There are also bending tests, for which the sample experiences both tension and compression.)

SAQ 1.8

The tensile strength is found where the stress reaches a maximum; in this case about 460 MN m^{-2}.

The 0.2% proof stress is estimated from where the line which is started from 0.2% strain crosses the test data; in this case at around 410 MN m^{-2}.

SAQ 1.9

1   No, it is completely generic, and designed to be applicable to any supplier–customer relationship.

2   The standard applies equally to both sides of the relationship (though different sections of the standard do have different emphases).

SAQ 1.10

1   The standard covers far more than testing, despite its title. For example, the following extract seems to me to be about the requirement that someone in the supplier company be responsible for the quality assurance process:

> The supplier's management with executive responsibility shall define and document its policy for quality, including objectives for quality and its commitment to quality.

Items (e) and (f) seem to me to be about the need for proper documentation, as is this:

> The supplier shall establish and maintain documented procedures for final inspection and testing activities in order to verify that the specified requirements for finished product are met. The required final inspection and testing, and the records to be established, shall be detailed in the quality plan or documented procedures.

And the whole standard could be said to be concerned with the need for a well-developed method by which the quality-assurance process is carried out.

2   This section outlines the principles which need to be adhered to for making measurements such as those to ensure the quality of eye protectors. In particular, it emphasizes the need for proper calibration of the test machines, to ensure that the measurements are made to an acceptable level of accuracy.

Part 2
Patents: The Engineer as Innovator

Contents

1 Introduction

In the first part of this block you were introduced to standards as a way that engineers communicate methodology and best practice. However, in many cases, engineers want to ensure that any new development they make is protected from unscrupulous exploitation. A new product, a new mechanism or a new technique can be extremely valuable in the marketplace, and there would be little incentive for invention and innovation if a new development could instantly be copied and marketed by anyone with the manufacturing capability.

Hence the existence of ▼Patents▲. In this part of the block we will look at the system of patents, in the context of how they are developed and how they can be used to protect an idea or a product from infringement.

Product invention and innovation is taken for granted by many, as something which 'just happens'; perhaps just as a natural part of the design process. It is also often assumed that 'invention' is the prerogative of the scientist rather than the engineer or designer; but a new discovery must be developed into a working machine or device capable of manufacture before the invention can be formally patented: a new theory or scientific observation may *lead* to an invention, but it is not sufficient in itself for the award of a patent.

Invention is not restricted to machines or devices, but also includes new materials, compounds (such as drugs), and processes. In this part of Block 3, we will concentrate on simple mechanical devices. It may be surprising, but it is true that simple devices are still being invented to solve fairly mundane problems, such as hollow beams to support brickwork or plastic shells to encase lawnmower motors. The same inventive principles underlie simple devices as the much more complex machines which often attract public attention, and which tend to obscure the basic simplicity of the inventive process.

Invention is closely linked to the design process. We discussed in Block 2 that invention is often the starting point in generating a successful design. Patenting is just one stage in the process: a patent is not a guarantee of making a successful product; you will see that neither is it necessarily a guarantee of having made a genuine invention.

The general aims of this part of the course are as follows.

▼Patents▲

The patent system as we use it today dates back to Elizabethan and Jacobean England, when the government wanted to provide a monopoly for innovators. But the idea of patents is much older, and goes back to at least the reigns of Edward II and III.

Such for example is the grant in 1315 to the town of Worstead in Norfolk for the manufacture of worsted cloth. The remit of the patent was at that time therefore much wider than just newly invented products (worsted being widely known before the grant of a patent), and extended to whole branches of industry (such as mining). But the system could then be abused by monopolists imposing artificially high prices in the total absence of competition: a recurring feature of any system that grants sole rights to one person or company. In the modern patent system, a patent expires 20 years after it is granted, so any monopoly can only be short term.

From Georgian times, the document granting a patent required the inclusion of a description of the invention, and the way it worked. The first recognizably 'modern' patent dates back to 1711, granted to one John Nasmith, who had an idea for preparing and fermenting the wash from sugar and molasses.

Outside the UK, patent systems which were not dissimilar were developed in other European countries, and in the United States (which was especially keen after the 1789 War of Independence to remove any royal connection with the patent process!). One important difference remains between the USA and the rest of the world, however. The US patent system stipulates that the 'first to invent' wins the race for patent protection, rather than the 'first to file' principle adopted virtually everywhere else in the world. Inventors seeking patent protection in the US will therefore have to keep extensive records of the steps which led to the invention. Engineers and scientists working in companies use notebooks with pages that are numbered, dated, and countersigned by a manager, to show the progress of their development work.

1.1 Aims

- To encourage an appreciation of the importance of product invention and innovation for practising engineers.

- To develop an understanding of how product design may be protected from copying.

- To show how inventive ideas are used in product development.

- To show how the design process works in the incremental development of products.

- To show the importance of precise definitions in the description of inventions for patenting purposes.

1.2 Problems in collieries

Before we look at any devices in detail, it is useful to try to examine what invention involves. Invention usually starts with a problem, and it is the solution to that problem to which the invention is addressed.

The original miners' lamp was designed to be the solution to the problem of explosions in coalmines. Until the invention of this lamp, mines were lit by candles. This was very dangerous because the open flame could ignite the methane gas (or 'firedamp') found in coal pits. This methane explosion could then cause a larger explosion in the clouds of coal dust disturbed by the first blast. A disaster in 1812 at the Spedding colliery in County Durham killed over 50 miners, and spurred many people, including Humphrey Davy, to seek a safe light source.

Davy solved this particular problem by reasoning that what was needed was a way to restrict the flame inside a shield, so that it could not ignite any methane outside of the lamp. However, sealing the flame completely in a enclosure would just cause it to go out through lack of oxygen. Davy reasoned that a metal gauze (a type of mesh) could prevent a flame escaping to cause an explosion. But what size gauze to use? If the gauze mesh is too coarse, the flame will pass through the gaps. Too fine, and there will be a reduction in the light output. In fact what Davy did was to show that a minimum gap size is needed in the gauze, of around 1 mm, a value that he determined by experiment.

Davy never patented the invention. Ironically, his important contribution (shown in Figure 2.1) led to greater numbers of explosions in the short term. The reason was that the iron gauze rusted very quickly, and the critical mesh size was exceeded if only one interconnection was broken. Later inventors in the middle of the nineteenth century created better versions of the lamp by using several gauzes (the idea of fail-safe, so that if one rusted, there were others to prevent the flame penetrating), and by using glass to surround the flame below the gauzes.

Until very recently, lamps of Davy's basic design were still in regular use for testing of gaseous atmospheres. If methane gas is present, it will burn as a blue cone above the main flame (with the gauze preventing the flame from igniting the methane in the mine), and the height of the cone enables the estimation of the concentration of methane in the air, from 0.5% up to several per cent.

The mine workings must be evacuated if the methane concentration reaches 1.25%, so the lamp gives critical warning of dangerous conditions. It has now been replaced by methanometers, which monitor the methane concentration using an alumina sensor soaked with palladium or platinum catalysts. However, the sensor is easily poisoned by sulphur, which is prevalent in

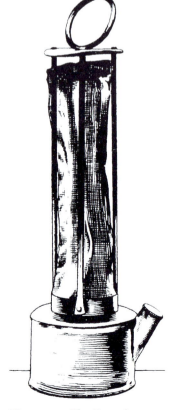

Figure 2.1 The Davy lamp

mines, so the devices need constant and regular maintenance, and there is still much reliance on flame lamps for their ease of use and reliability.

SAQ 2.1 (Learning outcomes 2.1 and 2.2)

(a) Describe briefly the problem which led to the invention of the flame safety lamp.

(b) What single key idea led to a solution to the problem?

(c) How was this concept translated into a working device?

(d) Indicate what practical problems arose when the device was used in collieries, and how those problems were overcome.

1.3 Lighting inventions

The Davy lamp is just one example of an inventive solution to a problem. In that case, existing materials were used to provide the answer. In some cases, though, the development or use of new materials is part of the inventive process, and the new material allows the creation of new products. In this section I will look at some lighting inventions that progressed through materials development.

Although flame safety lamps were only of limited applicability outside coal mines, there was a long-felt need for better lighting during much of the nineteenth century. By 'better', I mean more intense: a brighter light. With the growth in manufacturing industry, complex machines needed continuous supervision and maintenance, and illumination by candles or oil lamps was far from adequate (even during daylight hours). Although intense sources of light had been known since Faraday's work on ▼The carbon-arc lamp▲, they were not readily portable and were not safe or cheap enough for regular use either in industry or in homes.

▼The carbon-arc lamp▲

The carbon-arc lamp was one of the first light sources to exploit the then new discovery of electricity. The device was fairly simple: a very high voltage was generated, and applied to two carbon rods. Holding these rods close together produces a spark between them (Figure 2.2) because a high *electric field* is produced in the air gap between the ends of the rods. With a powerful electricity source, the spark will be continuous, and can be extremely bright.

The problem with this design is that the rods are burnt off rapidly, and so need to be continuously pushed together in order to maintain the electric current and hence the light. If the gap between the rods becomes too large, the electric field will drop to a point where the spark will no longer form.

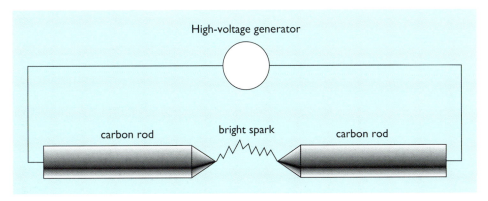

Figure 2.2 Principle of the carbon-arc lamp

The development of coal gas – flammable gases extracted from coal – provided a readily available source of energy when piped distribution systems were constructed. However, the illumination from naked flames was not great, being not much better than a candle, and there was the continuing danger from fire. This problem did result in the invention of the incandescent mantle, where the flame is played on a woven structure impregnated with chemicals which shine brightly when heated in an open flame. A similar development used lime as the incandescent material (hence the expression 'limelight').

One problem with all these methods is that a high temperature is generated by the light source, which could clearly be a fire hazard. There was no 'safe' lighting system available.

Although the concept of electricity had been understood for some time, the development of electricity-generating systems (pioneered by Edison in the USA) facilitated efforts to develop a light source powered by an electric current. If this could be done successfully, then electric light could be the solution to low intensity (candles, oil lamps) or dangerous (carbon arc, gas flame) lighting. Although electric lighting works by heating a wire until it becomes hot enough to glow brightly (it becomes *incandescent*), the invention of the light bulb neatly sidesteps the danger of having exposed high temperatures by enclosing the incandescent source in a glass envelope. The key to the solution of the problem of electric lighting lay not so much in the need for an electric current, since this was already available, but rather in the need to find a material which could withstand the high temperatures needed to reach incandescence and glow brightly without melting; which would conduct electricity; and which would last for a reasonable time.

The solution was found independently by Edison in the USA and Swan in Britain in 1879: heated carbonized filaments would provide a *continuous* light if they were protected from oxidation (i.e., burning) by being enclosed in either an inert atmosphere (which does not react with the filament) or a vacuum (Figure 2.3). Both solutions worked, although the lifetime of the lamps was still limited by today's standards. Nowadays, the preferred filament material is tungsten. It possesses a very high melting point (3410 °C) and can be readily fabricated into the coiled wire filament of electric light bulbs.

Figure 2.3 An early carbon – filament lamp

SAQ 2.2 (Learning outcomes 2.1 and 2.3)

(a) What was the key problem in making the first incandescent electric light?

(b) What had been the previous methods used for producing light artificially? What were the perceived problems with these methods?

(c) Why was choice of material important for the electric light filament?

1.4 New materials

Many problems have been solved by the application of a new material or process. The unique properties of a new material can make possible products or product forms that were previously unthought of. This is a theme that I will explore further in introducing the steps of invention and product development.

The development of the carbon filament seemed a logical step, because it had been shown very early in the same century by Davy and Faraday that the illumination from a candle is supplied by brightly-glowing, heated carbon particles. (The experiment is easily carried out: simply play a flame against a cold surface and the surface will quickly blacken with the deposit of carbon particles from the flame.) Inventors already had some idea of what direction to proceed since it was known that carbon could conduct electricity (hence its use in the carbon-arc light). Thus Edison performed experiments on thousands of carbonaceous (carbon-rich) materials before arriving at bamboo, which could be charred to leave a carbon filament that would conduct electricity.

This trial-and-error approach to solving a problem has always been one of the principal ways in which inventions are made, and is still an important tool in the inventive method. It is not a step entirely into the unknown, however, because the inventor will have a good idea of the properties needed, and the classes of material which will meet those needs: the inventor builds on what has gone before.

For simple products, many materials may be acceptable, especially where the desired main property criteria are mechanical (like strength or stiffness) rather than for more specialized applications (like optical or magnetic properties) – we have already discussed strength and stiffness: another important mechanical property is ▼**Toughness**▲. Thus a chair can be made in steel, wood, plastic, composites, textiles, and even inflated rubber.

▼Toughness▲

Put simply, toughness is just the opposite of brittleness. A material is said to be brittle if it is easily broken by an impact. Most ceramics are brittle: you will be aware that a ceramic mug will shatter if dropped onto a hard surface. Metals, on the other hand, are almost always tough materials: a dropped metal saucepan might suffer a dent, but will not shatter.

This is essentially the difference between a brittle material and a tough one: tough materials will tend to absorb damage by denting or changing their shape permanently in some other way, whilst brittle ones will break.

You can demonstrate the difference between tough and brittle to yourself using a metal paperclip and a piece of chalk. You will find that bending the paperclip doesn't cause it to snap (if you bend it back and forth repeatedly you can break it, but this shows a different phenomenon – metal fatigue), even though a lot of deformation is involved. The chalk, on the other hand, snaps in two very easily, despite being thicker. The chalk (which is a ceramic material) is much less tough than the steel used to make the paperclip.

Toughness is not as easy to calculate for a material as strength. Strength just needs a measure of the force needed to break the material and the area of the sample over which the force was acting. Calculating toughness requires a knowledge of the force required to break a specimen of the material that has a crack of known length in it. Figure 2.4 shows a picture of a typical specimen used to test the toughness of metal samples, and how it is loaded. A crack is grown into the sample by imposing metal fatigue on it, which involves loading and unloading the sample many tens of thousands of times, at a load much below that at which the sample would normally fail. This causes a crack to grow in the sample, and the geometry and loading of the testpiece are designed so the crack is grown in a controlled manner.

The sample is then loaded gradually until it fails: the load used to fail the sample will be much higher than that used in the previous step to induce a crack. The toughness is calculated from knowledge of the load, the

crack length, and the geometry of the sample. The units of toughness are not as easy to understand as those for stress: they are N/m$^{3/2}$ or N m$^{-3/2}$. This is more complicated than the calculation of stress, because the measurement of toughness is more dependent on the shape of the sample: for the testpiece shown in Figure 2.4, the toughness K is calculated from:

$$K = \frac{Y\,P}{B\sqrt{W}}$$

where B and W are the sample thickness and width respectively, and Y is a number (with no dimensions) that is a function of the sample geometry and the length of the fatigue crack when the sample is broken. P is the load at which the sample fails.

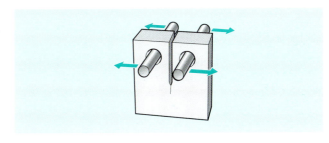

Figure 2.4 A sample used to measure toughness

You may ask yourself: how did my simple paperclip and chalk test for toughness manage to show anything, as there were no cracks visible in my samples? The answer is that all materials contain small flaws or cracks, even if they are not apparent to the eye: they may be only fractions of a millimetre in size. Toughness is a measure of how susceptible a material is to fracture, even if obvious large flaws are not present.

Toughness is very useful for designers, because it can be used to predict whether small flaws in a structure are safe or unsafe. Good toughness is important if you're designing a bicycle which must be reasonably resistant to heavy bumps, or a knife that won't splinter in the washing-up.

Each material has its own characteristic set of properties, and within each of the very large classes of material, there are numerous types and grades which can show enormous variation of individual properties. Iron, for example, can be obtained in several forms: as wrought iron (soft and ductile), steel (tough), and cast iron (brittle); wood can vary from a stiff and heavy hardwood like oak to softwood, and even includes a light foam (balsa wood). Aluminium is much lighter than steel but can meet and achieve equivalent functions, although at greater raw material cost; modern plastic materials can offer a wide variation in properties.

As the property specification becomes more severe for a specific product, the choice of materials diminishes. Let's look at two examples of how materials are used to meet a product specification.

Kitchen table

A domestic kitchen table has to meet several functions: act as a level worktop for the preparation of food for cooking, support moderate loads when in use either as a work bench or as a surface from which meals can be eaten; and be resistant to impact blows from users. It should also resist the moderate heat and humidity which arises during cooking and related tasks, and might have to withstand the heat of a hot pan from the cooker being placed upon it. An additional requirement might include wear resistance for the worktop itself, so that knife blades do not scar or harm the material of construction. Such scars could act as a reservoir for food debris, which could decay there and form a breeding ground for harmful bacteria. These are all functional requirements of the design. There are also constraints on the form: the dimensions should conform to the space available in the kitchen and the height to the normal working height of the cook.

The materials used must therefore be capable of withstanding loads applied to the working surface without noticeable distortion, be impact and wear resistant, and resist the effects of moisture and heat. Many woods such as oak and pine are well capable of meeting these demands, and are the 'traditional' materials of choice. Modern equivalents include combinations of steel legs and plastic-faced wooden worktops, where the plastic finish may offer greater wear resistance and hence be less likely to act as a source of bacterial contamination, but may not be suitable for the placing of hot objects.

Light bulb

The outer case of an electric light bulb presents a much tighter property specification than the table, since all the materials of its construction must be able to withstand the high working temperatures produced by the incandescent filament. The exterior shell (or envelope) of the bulb, which is the prime container for the glowing filament, must also resist atmospheric pressure since it encloses a vacuum around the filament itself. It must be translucent or transparent to allow light from the filament to escape. The base to which the envelope is attached needs to provide safe electrical contacts for the filament. The material used in the bulb must also be corrosion resistant.

Glass is the only material capable of meeting the stringent requirements of mechanical and thermal stability needed of the envelope as well as being transparent to light. Its thickness can be controlled to provide varying levels of mechanical resistance (to external impacts for example), although a domestic bulb is normally very brittle. Metal contacts are usually made from aluminium sealed in an insulating material, and the cap is also normally made from aluminium.

SAQ 2.3 (Learning outcome 2.4)

(a) What is the function of a lightweight ladder with 15 rungs for use around the home?

(b) What materials properties should be included in its specification?

(c) Describe from your own knowledge what materials are used in such a product and how it meets the specification.

(The answer to this SAQ includes mention of ▼Angles▲, so it would be helpful to read this input first.)

▼Angles▲

Quite often the relative position of two objects is defined in terms of the angle between them. A ladder leaning against a wall might carry a warning that it should be used at an angle of around 75° (75 degrees) to the horizontal (Figure 2.5). Note that it's important to say what the angle is relative to: the angle that the ladder makes with the wall is much smaller: 15 degrees.

By convention, a complete circle sweeps out 360 degrees. A ladder standing vertically would be at 90° (a quarter of 360°). You will probably already have a familiarity with this notation for describing angles.

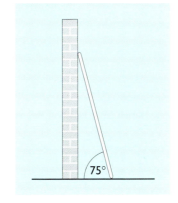

Figure 2.5 A ladder leaning against a wall

1.4.1 What is an 'invention'?

The range of materials available is currently expanding at a high rate, giving designers yet more choice for a specific product. When a new material is introduced, inventive minds immediately examine the material for its new properties. This method is not new, and can be illustrated by the introduction of rubber from the New World in the 18th century. It had been used by native Amerindians as a material for making pots and shoes, as well as for leisure activities such as ball games (see ▼The discovery of rubber▲).

▼The discovery of rubber▲

Spain's desire to colonize South and Central America was driven by the greed for gold and silver. But of the many new discoveries made, one plant was destined to prove extremely valuable in the long term. It was *Hevea brasiliensis*, a large tree which exuded a latex sap when cut. When collected, the sap could be dried to a solid material and fabricated into many useful products. It was seen by Christopher Columbus himself during his second voyage between 1493 and 1496, who saw Amerindians playing a game with rubber balls. This game had been developed by Olmec Indians as early as 1000 BC into an extraordinary sport known as the Sacred Ball Game. It was played only by the most important members of society in a rectangular court 283 feet long, 100 feet wide and 27 feet high (shown in Figure 2.6).

After the solid ball was thrown into the court, the players had to pass it to team mates using hips, legs or elbows, and manoeuvre the ball into one of two rings in the centre of the court to score. Side bets were usual, and the winners had the right to strip spectators of clothing and jewellery. Omens were read from the way the game developed, as well as from the nature of the victory.

Apparently the losing side could pay with their heads (as the drawing suggests): there appear to be many parallels with modern soccer!

Figure 2.6 The Sacred Ball Game

It was formed by collecting liquid latex sap from one species of tree, removing the water in it (by boiling it away) and allowing the semi-fluid resin product to harden over shaped formers by heating. The product had substantial elasticity, capable of being extended to many times its original length (think of the extent to which you can stretch a rubber band). This elasticity also means that items made of rubber show tremendous bounce! No other materials behaved in this way, so it naturally became the object of enquiry when it reached Britain in around 1750.

One of the first applications for rubber happened almost by chance. Blocks of the material when rubbed against pencil marks were found to remove them. Such a simple use would not be expected from a planned research effort! This is an example of invention by accident or happenstance, and in this case, the word 'rubber' became absorbed into the language as a term for an eraser.

What were the main features of this 'invention'? They were:

1 the discovery was new (no one had done it before, apparently);

2 the result was unexpected (not easily predicted from existing knowledge in that particular field) and not at all obvious to workers in the field (or *art*);

3 the effect could be applied to making a product for use by many people (it is capable of manufacture).

Moreover, once the discovery had been made, systematic research led to better erasers. It was found later that adding various fillers such as chalk will improve the effect. The filler actually weakens the rubber material, making it easier for pieces to be abraded away, and so taking the graphite smears of writing away more easily. So perhaps we ought to add another feature:

4 the basic invention was capable of improvement (better versions can be made).

The first three points are in fact incorporated into any legal framework for defining new inventions: they must be

1 novel,

2 original (non-obvious), and

3 capable of manufacture (or otherwise realized in a functional way: not all patents cover mechanical devices).

SAQ 2.4 (Learning outcome 2.5)

Does the development of the safety lamp by Humphrey Davy qualify as inventive according to the three main points given above? Justify your answer against each of the three points.

Recall our definition of an *invention* from Block 2: a novel idea that has been transformed into reality – given a physical form such as a description, sketch or model conveying the essential principles of a new product, process or system.

The three points above are capable of application to any new product, whether inventive or not. Lawyers have codified them and developed specific legal tests to ensure that a new product can be evaluated for its originality. We have already come across the fourth criterion in the discussion of flame-safety lamps. They show evolutionary development from the original Davy Lamp, in addressing problems of integrity and longevity in the severe environment of the mine. Each individual improvement is patentable if it meets the three basic criteria of novelty, originality and manufacture. We have said already that there must be some context of need, also: there must be some use for the invention.

The basic test used to evaluate any new material discovery is therefore to ask three questions:

1   is the application novel?

2   is the use original, or not obvious to a skilled practitioner?

3   can the material be manufactured for this specific application?

If it has not been used at all anywhere, then the application is novel, but if some record is found, or a witness comes forward who can attest the use of the same material in the same application, then it will not be novel, and the application cannot be inventive.

The second question is much more difficult to answer, and will depend on the skilled practitioner who responds. If a not dissimilar material had been used before in such an application, then it is conceivable that using the new material could well be obvious.

The final question is the most straightforward to answer: if the material is capable of manufacture, then it satisfies the third criterion.

1.4.2 Innovation in processing

An example of development and innovation in processing is the story of iron and steel. Although wrought iron was well known for centuries (since the so-called Iron Age), it needed heavy working by blacksmiths in order to shape it into usable products. It was also rather soft and susceptible to corrosion (rusting). Harder products such as swords could be made by repeated working and folding, but steel as we know it today was unknown until the 1850s.

Large scale production of iron was only achieved relatively recently, by Abraham Darby in the 1750s (see ▼The Darby family and cast iron▲). So-called 'cast' iron became the structural basis of the first phase of the British Industrial Revolution, being widely used for large structures such as bridges (e.g. the Coalbrookdale bridge: Figure 2.7), buildings (e.g. The Crystal Palace: Figure 2.8) and for a host of applications in industrial machinery.

Figure 2.7 The Iron Bridge at Coalbrookdale

Figure 2.8 The Crystal Palace

▼The Darby family and cast iron▲

The development of cast iron as an engineering material is very much the story of the Darby family, who developed large-scale methods of making this valuable material as we saw in Block 1.

The key step in smelting iron ore to raw metal is the *reducing agent* used: a reducing agent is a chemical that reacts with the iron oxides in the ore to release the iron in metallic form. Charcoal (produced by partial burning of wood) had been used for two millennia for making wrought iron in small quantities, but it was the development of the coal industry which prompted Abraham Darby to try to use coke (made by controlled heating of coal in the absence of air) as a more effective fuel than charcoal. This development in 1708 led to the cast iron industry founded on the banks of the Severn at Coalbrookdale. Coal itself cannot be used for the ironmaking process, owing to impurities such as sulphur which impair the qualities of the iron produced, and it was coke which was the key step in developing a furnace capable of making cast iron on a large scale. The Old Furnace (shown in Figure 2.9) was the forerunner of the modern blast furnace, and was used to make the members of the first cast iron bridge, spanning the Severn at Coalbrookdale (Figure 2.7).

Coke, together with limestone and iron ore, was fed in at the top and heated by burning in air fed in lower down (not shown); the molten cast iron was extracted at the base. The air was fed in by *tuyeres* – pipes leading in about halfway up the furnace, which 'blasted' a draught of hot air to the charge. The mechanical properties of the coke were important because the mixture had to be porous enough so that reduction proceeded smoothly, and it had to resist the weight of material above. The molten iron could then be tapped and run directly into moulds. This furnace was especially important for making the key parts of steam engines. Some of the carbon of the coke dissolved in the iron (about 4%) to

Figure 2.9 The Old Furnace

give the material its relatively low melting point, but also made the material rather brittle.

This was a key discovery: that the amount of carbon present in the iron controlled not only its melting point but also its properties. By controlling the additions of carbon through the use of coke, a form of iron was made which could be cast on an industrial scale. When we talk of *iron* as the material used for a range of engineering use, we are almost always referring to an alloy of iron that contains some carbon.

1.5 Limitations of new materials

Quite often a new invention is found to have deficiencies when it is tested in use. Rubber, for example, when its properties were examined in more detail, was found to deform slowly under an applied load, a phenomenon known as ▼Creep▲. Another drawback to the use of rubber in products was its tendency to melt and become sticky at even moderate temperatures, now known to be related not so much to the purely physical attribute of melting, but rather to chemical degradation. The material, in other words, oxidized slowly in air.

Of course this was just another set of problems waiting to be solved. It was discovered – almost accidentally, by Goodyear – in 1846 that rubber could be stabilized by heating with sulphur, in a process called *vulcanization*.

As with any fundamental advance in technology, the initial discovery of a new material is followed by a sequence of further discoveries which widen the scope of the original invention or discovery, and any associated patents are known as 'improvement' patents. It is a stepwise sequence, each further step relying on the previous development. What came before a particular invention is known as the 'prior art' – the accepted knowledge base – of this particular area of invention.

▼Creep▲

Creep is a phenomenon whereby a material will deform very slowly as a result of a stress. Application of a load, or stress, causes a strain to occur, as we saw in the previous part of this block. When creep occurs, the strain continues to increase over time, and will no longer return to zero when the load is removed.

The effect of creep can be observed in lead pipes (used in older buildings) which have gradually sagged over the years. In more modern materials, plastic guttering can also show sagging between supports. Creep is accelerated as the temperature rises, so it is very important in engineering applications where the temperatures are high, and for metals, where they are being used close to their melting point. So materials for jet engine combustors need to be creep resistant, but so also do the alloys used for soldering electronic components on circuit boards: because they have low melting points for ease of manufacture, they can be susceptible to creep at ordinary room temperature (we will revisit this example later in the block).

Problems in exploitation were also encountered with cast iron: it is brittle in tension, like stone. The design of structures using cast iron must allow for this deficiency, as with stone. But one important advantage of cast iron is that it can be made into large beams, rather like wood: it does not have to be used in blocks like stone. So the designer of the Coalbrookdale bridge made the cast beams to fit together like wooden beams: the ends were shaped so that they could mesh together just like the joints on a wooden structure, and the joints were pinned with dowels.

1.6 Development

1.6.1 Rubber and polymers

Following the discovery of any new material, there will be further developments in the process technology needed to make new products. Thus even prior to the discovery of vulcanization, attempts were being made to use rubber in products such as wellington boots and mackintoshes by coating fabric with rubber. In order to achieve this, it was necessary to develop machinery capable of working with rubber on a commercial scale. Hancock, working in Manchester, developed large machines to 'masticate' the raw bales of rubber into a material which could be worked more easily into finished products. It was found that the same process also made the rubber soluble in organic solvents such as naphtha, so a solution of rubber could be applied to fabric, for example to make a coated cloth or 'rubberized' fabric. Together with Goodyear's discovery of vulcanization, it made natural rubber into a material capable of being mass manufactured into a variety of new products.

One of the products into which very large amounts of rubber would eventually be fabricated was the familiar pneumatic tyre (see ▼Thompson, Dunlop and the tyre▲).

The process of vulcanization can also be used to make rubber into a rigid and quite inflexible material called ebonite (loosely named after the hardwood ebony). Such a state is similar to that for many natural plastic materials.

Natural plastic materials have been well known for millennia, the oldest perhaps being amber, which was traded widely in prehistoric times. Amber's value lay not just as a decorative jewel, but also as an adhesive for fixing stone tools to wooden hafts. Amber is just fossilized resin exuded from a coniferous tree, and just like rubber, is based on molecules of carbon atoms linked together to form chains (polymers). There are numerous other natural resins, such as lac, the exudate from a certain species of beetle found in the Far East. This was imported into England by merchants and rapidly exploited for making hard 'lacquers' for furniture. Horn was widely used for making spoons, and even beaten flat to make thin transparent sheet for lantern windows. However, the first experiments on natural plastics were made by

▼Thompson, Dunlop and the tyre▲

One inventor, Thompson, in the 1840s had the bright idea of improving the suspension of carriages by providing them with rubber tyres inflated by air. Previously, solid wooden-rimmed wheels were the norm. The problem of rough journeys was a long standing one on what then passed as roads, and the reason for this can still seen on unclassified country lanes or byways. Such roads were rarely provided with a foundation of ballast or stone, except perhaps at the most vulnerable points at stream crossings. They were pot-holed and worn into an irregular surface, so normal passage was difficult and uncomfortable to passengers in horse-driven carriages. Large diameter wheels with a springy suspension could alleviate the effects of some of the worst holes, but unpleasant vibration from any irregular road surface was inevitable.

So, why not make a wheel from a material which is inherently capable of absorbing vibrations? And then make it even more vibration resistant by creating a cushion of air? The invention of the pneumatic tyre by Thompson in 1846 seemed to solve the problem, but the idea never caught on with carriage makers. It was not until much later, in 1888, that the rediscovery of the principle was made by a Belfast vet, John Boyd Dunlop.

The motive for Dunlop's invention was an appeal from his son (aged 10) to improve the solid rubber tyres on his tricycle, especially for when he rode on the granite setts and tramlines of the Belfast streets. John Dunlop's first experiment was conducted on a solid wooden wheel, to the edge of which he attached a rubber inner tube, protected with an outer sheet tacked to the rim. The inner tube was a simple modification of a football inner diaphragm. On inflation, comparison in a rolling experiment with the solid tyre showed that the pneumatic tyre was greatly superior in performance. (His work as a vet helped in the construction of this and later modifications, because he apparently made and used various rubber products for his veterinary work.) The first manufactured tyre involved wrapping the outer sheet over the rim and between the spokes (Figure 2.10)

Dunlop's first patent was later declared obvious (and hence invalid) in the light of the earlier specification of

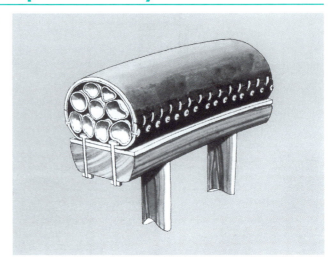

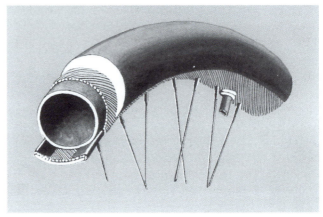

Figure 2.10 Thompson's tyre of 1846 (top) compared with Dunlop's first tyre (bottom)

Thompson, but it didn't deter Dunlop from patenting various modifications to the prototypes, such as a valve for the inner tube. Other inventors, Welch and Bartlett, developed a wire bead so that the tyre could be attached freely to the steel rim of the spoked wheel, the patent being bought by Dunlop for the company he formed to make the new bike tyres.

Parkes in the middle of the 1800s in an attempt to make billiard balls to replace the expensive ivory balls then in use. The material was called Parkesine, and made using a derivative of cellulose. So-called 'stuffing machines' were developed (Figure 2.11) to make the product, a forerunner of the modern method of injection moulding.

An improved plastic material, called Bakelite after the inventor Dr Baekeland, was introduced in 1877. It was made from by-products of coal tar, itself a by-product of the manufacture of coke from coal. Two products in particular, phenol and formaldehyde, could be polymerized together to form a hard material. The material, still used today, is called a *thermosetting* plastic, because it cannot be reheated to become fluid again after it has been processed because the material has effectively become a single giant molecule: the atomic chains in the polymer are joined together by chemical cross-links.

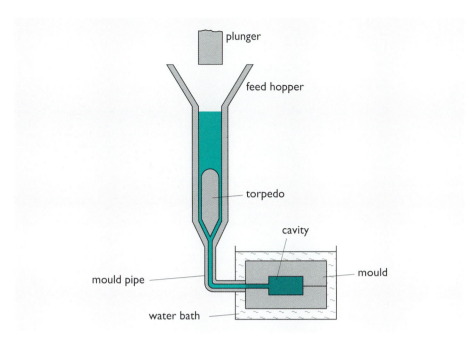

Figure 2.11 The stuffing machine of Hyatt

The major *thermoplastic* polymers were introduced much more recently, polyethylene only being invented in the 1930s: a *thermoplastic* polymer can be reheated to soften and/or liquefy it. A large number of thermoplastics have been developed since the 1940s, with a wide range of properties and applications. Many other plastics have and continue to be made synthetically, and their exploitation in engineering products is a vibrant and expanding area of innovation. They offer greater resistance to the environment than many conventional materials, and manufacture by processing to shape is relatively easy.

1.6.2 Cast iron and steel

The development of steel from cast iron proceeded not so much by mimicking natural materials, but by fundamental advances in manufacturing methods; especially when it was realized that the key to understanding the importance of steel was a knowledge of the role which carbon plays in the metallurgy of iron. Cast iron, as mentioned earlier, is brittle because it has a high carbon content (about 4%).

More ductile 'wrought' iron could be made, but only by a slow, small-scale and labour-intensive process, so iron was an expensive commodity. Since wrought iron (which has a very low carbon content) and steel (which typically has below about 0.4% carbon) were essential materials for making tools and structural products such as railway lines, a way had to be found of making iron and steel on a large scale.

The critical step was made by Bessemer in 1856, in a series of classic experiments with various designs of furnace. At one point in his work, he suddenly realized that he didn't need to heat or supply fuel to the charge of molten pig (cast) iron when trying to make steel: the 4% carbon present in the pig iron would burn, and produce heat, if an air stream was directed through the molten metal, so keeping the metal hot and fluid *as well as reducing the carbon content*! The result was the Bessemer furnace (shown in Figure 2.12).

The furnace was designed with tuyeres (those pipes again) at the base through which air could be blown, and through which the iron or steel could be tipped out at the end of blowing.

FIG. 43. THE FIRST FORM OF BESSEMER MOVEABLE CONVERTER AND LADLE

Figure 2.12 The Bessemer furnace

After selling (expensive) licences to clamouring iron masters from all over the country, all initial trials were disastrous. The problem was one of chemistry: the other iron producers used ore contaminated with phosphorus, which Bessemer realized by careful chemical analysis of the ores and pig irons (after the event) prevented the production of high quality steel. In his original experiments, he had fortunately used uncontaminated pig iron. As a result he set up his own steel works in Sheffield, but persuaded his suppliers to ensure the purity of the feedstock. The problem with the phosphorus-containing ores was solved by changing the lining of the furnace, the chemistry of which caused the phosphorus to be removed from the steel in the slag. The rest, as they say, is history.

2 Building technology: from concept to patent

In this section I will look at our first case study of invention: an example of where an inventive idea was applied to produce a new and successful product. The case study also shows the benefits of the patent system, as the company responsible for the invention was able to successfully win a legal battle against a competitor that tried to infringe its product.

We have in the previous section examined the way in which new materials have stimulated a cascade of new products using the unique properties of those materials in solving particular problems. The cycle of development will occur in any field of endeavour, provided there is a market for the new products and the benefits of the new products are recognized widely. This is not always easy, because there are still many industrial sectors where there is a reluctance to move away from traditional methods which have been tried and tested over many years, apparently with success.

For example, the domestic housing market still builds using traditional materials such as brick, mortar and concrete. Although some new materials have been widely adopted since the 1940s (e.g., breeze blocks, insulated blocks, fibreglass insulation, polyethylene film for damp-proofing, PVC plastic for rainwater goods and wastewater which are much lighter than the iron or steel equivalents, polyethylene pipe for water and gas, etc.), the standard brick plus wood for internal floors and roofs is the staple product for domestic structures. The components are of standard size so manufacturing costs are low. Various experiments with steel and/or concrete have not generally been successful, although small scale improvements have been made in narrow areas of application.

2.1 Lightweight lintels

I will take one simple example of the way in which new technology can enter this market – using traditional materials. I will show how some basic principles led to a patentable invention.

Bridging a gap has been a common problem for builders from prehistoric times onwards. Early examples are found in prehistoric tombs, such as Wayland's Smithy, a Stone Age long barrow in Oxfordshire. Stonehenge is a good example of large single stones spanning the gap between massive upright stones.

With walled buildings, smaller stones or bricks are used, and arches were developed for bearing the load above the gap, first widely exploited by Roman engineers (e.g., the Pont du Gard aqueduct) and then rediscovered by the Normans (Figure 2.13).

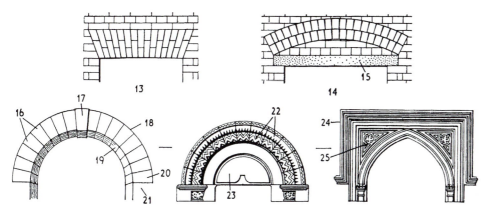

Figure 2.13 Historical development of lintels

As I mentioned in Block 1, a beam spanning a gap is put into bending by the weight placed upon it (which, of course, includes its own weight). There is a tension force on the lower surface, and a compressive force on the upper surface. The tension is a problem if the material used is poor in tension, like stone or concrete.

A hole in a building structure, like a door or a window, requires some form of bracing to prevent the brickwork from sagging into the gap, and possibly collapsing. A beam – called a lintel – is usually placed across the gap to provide the necessary support. Lintels for doors and windows were traditionally made from large beams of oak or even stone where this was available, which were of sufficient strength to support the imposed load from above. Such solid beams are, of course, very heavy.

With the development of domestic buildings after the First World War, steel-reinforced concrete lintels became popular because the material could resist the tensile force on the underside of the beam (Figure 2.14).

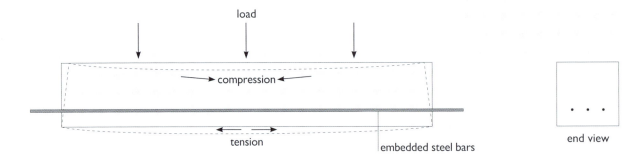

Figure 2.14 A simple steel-reinforced concrete beam

 Although the concrete is brittle in tension, much of the load on the tensile side is actually borne by the steel bars. A further development gives prestressed lintels, where the steel bars are tensioned before the concrete is laid and set. When the tensile stress is removed, the steel relaxes elastically (i.e., contracts back), and pulls the concrete into compression. This makes for a very much stronger material, because the compressive stress in the outer layer of concrete must be overcome before the bar is in tension and weakened. Using strengthening methods like this produces a beam that is, weight for weight, much stronger, so thinner and thus lighter beams can be used for lintels (Figure 2.15).

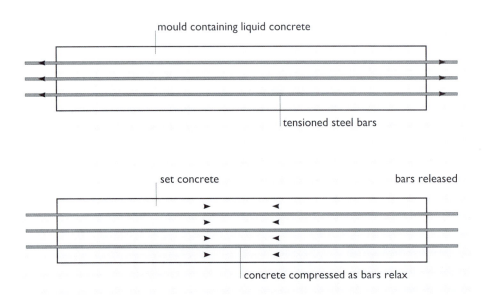

Figure 2.15 Producing a prestressed concrete beam

A prime factor in the development of many innovations in building technology is the ease with which component parts of the building can be manipulated and moved into their final position. Bricks are made in standard sizes, which keeps manufacturing costs down; cement and concrete are standard products. Wood is a traditional material which is supplied in standard sizes for internal joists and roofing timbers. Indeed, most roof timbers are now supplied prefabricated in standard shapes, so that they can be assembled quickly and the roof covering added soon after.

The importance of ease of assembly, and the availability of standard components, are factors central in the design of many products, either for particular domestic functions (e.g., vacuum cleaners, lawnmowers), or during the erection of large permanent structures such as buildings. Choosing the right components and assembly route are critical for successful manufacturing.

So what factors can inventors manipulate to aid this process? There are broadly two factors which are available to them: the materials of construction, and the design geometry.

2.2 Materials of construction

We have already seen that different materials possess different combinations of properties. The materials used in lintels are wood, stone, concrete and steel.

2.2.1 Material density

Each of these materials has a different density: a different mass per unit volume. This in turn is controlled by the atomic constitution of the material: how closely packed the atoms or molecules are in their structure, and the relative mass of the individual atoms in that structure (see ▼**Atoms and molecules▲**). The units of density, you should recall, are kilograms per cubic metre (kg m^{-3}).

▼Atoms and molecules▲

The original idea that matter could be composed of small discrete building blocks (as we saw in Block 1) dates back to Greek philosophers, but really only became a fundamental base of modern science in the 1800s. John Dalton, a Manchester chemist, reasoned that discrete blocks – atoms – must exist and can combine only in fixed ratios with one another to make compounds – molecules. All material can therefore be described in terms of a fixed number of different types of atom, which can occur as elements, or more commonly as compounds. So element A might be able to combine with element B in compounds with formula AB (one atom of A joins to one atom of B), A_2B (two atoms of A to each one of B – the subscript (the number below the line) relates to the preceding chemical symbol, to indicate its proportion in the molecule), AB_2 (two atoms of B to each atom of A) etc.

The way was then open to discover all the elements, and determine the exact formulae (composition) of pure compounds. Determining the formulae of materials such as steel or wood was more difficult because they usually consist of mixtures of many different compounds. Steel consists of a compound called cementite (whose formula is Fe_3C) dispersed in a matrix of nearly pure iron (Fe); some carbon (C) is also dissolved in the iron: the exact proportions will vary between different grades of steel. The properties of the steel depend on the distribution and proportion of the cementite, the crystal structure of the steel and so on.

Cement and concrete are still the subject of intense research into their precise constitution. Wood, as a natural material, is also complex in structure and constitution because the elements are bound together into long chains with other very large molecules. However, the long-chain compounds used in thermoplastics are relatively simple, and consist of identical repeating units or building blocks linked together to form chains. Thus polyethylene consists of chains of the repeating unit [CH_2-CH_2] linked together (recall this from Block 1). Other chain molecules (polymers) are more complex. ABS plastic contains three different types of polymer repeating unit in its structure.

 Some extra details about chemical symbols can be found in the SGSG, p. 128.

Exercise 2.1 (revision)

Calculate the mass of a rectangular lintel with a square section of length
1.2 m, thickness 9 cm and width 15 cm in the following materials.

(a) Concrete.

(b) Steel.

(c) Oak.

Take the density of steel to be 7800 kg m^{-3}, oak to be 1200 kg m^{-3}, and
concrete to be 2400 kg m^{-3}.

So the oak beam is the lightest, as one would expect from its low density
compared with the other materials. However, the concrete beam represents the
best value for money, being cheaper than oak by about five times; which
explains why it is widely available in builders' merchants. Solid steel is not
used at all in such applications, being far too heavy. It is possible to use it in
hollow sections, as we will see shortly.

The load-carrying abilities of each material are different. This depends on
several distinct and different properties of the material, including resistance
to movement under a bending load (related to the Young's modulus), long-
term resistance to load (whether the material deforms slowly – creeps – under
an applied load), and resistance to sudden failure from shock loading (related
to its toughness).

2.2.2 Load on a lintel

A lintel, by definition, supports the load imposed on it from the wall above.
The load puts the beam into bending. The load will not, in fact, be uniform,
because the interlocking of the bricks will tend to give some support, in a
similar way to an arch (Figure 2.16). The greatest load will be supported in the
centre of the lintel.

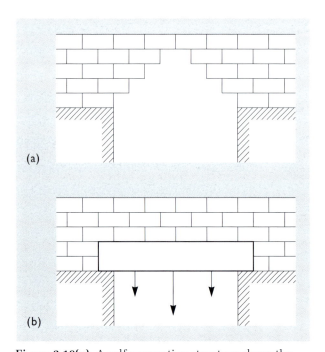

Figure 2.16(a) A self-supporting structure above the
gap **(b)** The load on a lintel

The load supported by any lintel is therefore rather modest owing to the self-
supporting nature of the brickwork. The lintel is not bearing the entire weight
of the wall above it.

2.2.3 Bending a beam

So what happens when a lintel is loaded?

Figure 2.17(a) shows a section of a beam along its length, with the strain exaggerated for emphasis. The material at the top surface is in compression, and at the lower surface it is in tension. Between the two surfaces, the material gradually changes from being in tension to compression, and it follows that there must be a crossover point where the stress is zero. This point is known as the *neutral point*. However, since the beam is bent continuously along its axis, there is a line of such points which represents the *neutral axis* (Figure 2.17(b)). (Beams are of course really three-dimensional, with a definite width, so the neutral axis is actually a neutral plane: an imaginary surface within the body where there is no load on the beam.)

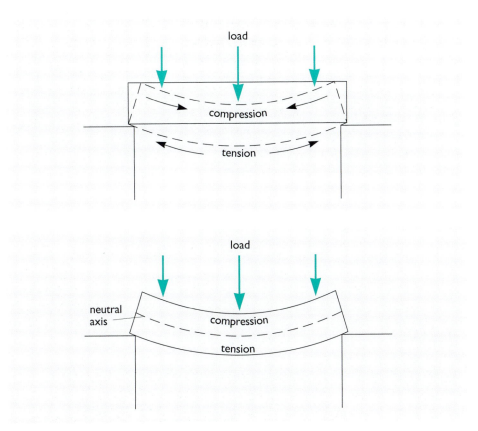

Figure 2.17(a) A lintel beam in bending **(b)** The neutral axis in a uniform bent beam is the line through a section of the beam where there is no stress on the material. Material present here is thus not used structurally

The realization that there is an inner part of the beam which is not loaded at all suggests an ingenious solution to the problem of weight-saving: simply remove the material from this region! It is performing no load-bearing role, so could be removed without a penalty to load-bearing performance or bending strength.

2.2.4 Hollow beams

By removing the centre part of a beam, substantial amounts of material – and so weight – can be saved, but how much? After all, it is only at the centre where the load is zero: even just off-centre, some load is present (either as compression or tension). If this off-centre material is removed, the material left will have to support more of the load. However, if the material is strong enough, this should not be a problem unless the material removal is taken to extremes.

Figure 2.18 Hollow concrete blocks wall

Figure 2.19 Bamboo scaffolding

Hollow beams in concrete or wood are really not feasible: concrete because it is a brittle material, and although hollow beams could be made easily enough from concrete, reducing the load-bearing ability of a brittle material is unwise; wood, although it is tough, would be expensive to make into hollow beams. In other circumstances, hollow concrete blocks can be made for decorative purposes, and even structural use for partition walls (Figure 2.18). Similarly, there are plant species which grow in a hollow form: the best example is bamboo. It is widely used in the Far East as an alternative to conventional metal scaffolding poles because it is readily available from local plantations (and is thus of low cost), is strong and lightweight. The strength is such that high-rise buildings can be built with this material (Figure 2.19). However, the circular form of the stem is not practical for use in lintels, and some other solution must be found for the problem.

2.2.5 Resistance to failure

Since concrete is brittle, it is rarely used in tension applications without reinforcing steel bars. But both oak and steel are tough materials which generally do not fail by cracking suddenly through the thickness. So if an oak lintel fails, there is likely to be some warning of failure, such as progressive sagging or slow cracking, rather than a sudden collapse. In fact oak beams in old buildings tend to creep with time, that is, deforming slowly under the weight of material without failure. Steel tends to fail after some plastic deformation (recall this from Part 1 of this block) following loading in excess of its yield strength, with buckling along specific lines depending on the loading conditions on the beam.

2.2.6 Environmental resistance

Another important property needed of a lintel is longevity: houses are built to last a considerable time, so any material of construction must resist attacks on its integrity from its environment.

Concrete is good in this regard, but steel and wood can be attacked by various agents. Steel in particular is susceptible to corrosion (rusting) and wood to attack from natural agents such as fungi (dry and wet rot) and boring insects.

The corrosion of steel occurs in the presence of water and oxygen. Corrosion is simply a chemical reaction where the iron reacts to form iron oxides: the powdery brown coating seen on rusted surfaces. The iron prefers to be tied into a chemical compound with oxygen rather than existing as 'pure' iron in the steel. Steel can be protected by applying a thin coat of zinc metal (galvanizing), or by polymer coatings. A polymer coating simply protects the steel from contact with the elements; a zinc coating has the added advantage that it will corrode in preference to the steel, giving an extra factor to the protection.

Wood can also be protected, with various chemicals which inhibit fungal attack and/or kill boring insects.

2.3 Design geometry

Other than by changing the material of construction, the main way in which an inventor can make a new lintel design is by way of altering the *geometry* of the beam. One of the problems with solid lintels, for example, is that they are very heavy (recall the answer to Exercise 2.1). That makes them awkward to transport to the building site, and awkward to handle and manoeuvre into position on the wall. As we saw previously, solid lintels could be considered to be over-designed for the job of supporting the superimposed load, so there seems to be an opportunity to devise some shape which will resist the load yet be much lighter.

2.3.1 Hollow steel beams

Steel can be fabricated easily into tubes or rectangular sections, so this would seem to be an ideal solution to saving weight. Sheet steel is a familiar product in various guises: cans for food and drink, car bodies, and so on. It should be straightforward to make hollow beam sections in sheet steel. Such sections are made routinely for ducting of hot air in ventilation systems, for example, simply by folding pre-cut flat sheet and joining it at the edges (Figure 2.20). This product supports very little load, so need only be thin (1–2 mm), but if the thickness were increased, it is clear that structural products could be made.

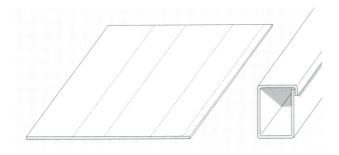

Figure 2.20 Fabricating a hollow steel beam

Fabrication of a product like this in thicker steel should not be too difficult, only requiring more robust presses to fold the thicker and thus stiffer sheet. Joining can be achieved in several ways: by folding over and clipping the sheets together; by welding; or by riveting. Moreover, the sheets can be so cut that an overlap is present at the joint, which can be used to mate with brickwork. This is the invention made by Catnic Ltd. in a patent of 1968, and this is now a staple product for builders of small dwellings (Figure 2.21).

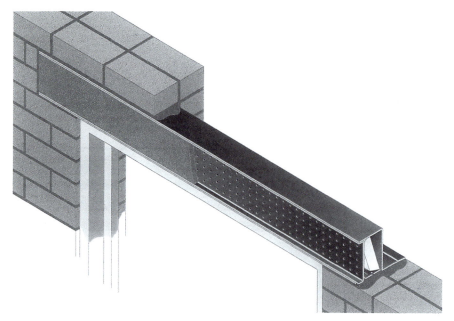

Figure 2.21 A hollow steel Catnic lintel

2.3.2 Shaped steel beams

We should consider other shaped beams which have lower weight than a
solid section, even though they may not be appropriate for use in lintels.
Shaped steel girders are one way of lowering the weight of a beam, and the
familiar I-section beam achieves the objective of providing support where it is
needed (Figure 2.22). It has less weight, obviously, than a solid bar, but is still
stiff enough for structural use (see ▼**The stiffness of beams**▲).

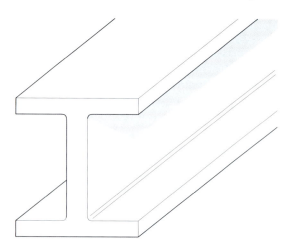

Figure 2.22 An I-section steel beam (I-beam)

The flat caps represent the top and bottom of the beam, where material is
needed and the cross-bar simply links the two surfaces together. Such beams
are of course widely used in the construction of large buildings, which are
made from I-beams welded or riveted together to form a superstructure within
which bricks or other small structural units can be used. Such beams can also
be used to form lintels, but usually are very much stiffer than is actually
needed. Although they are marketed in standard sizes, such sizes are too large
and rarely correspond exactly to the dimensions needed for small-scale
brickwork. The thickness of steel used is usually a minimum of about 5.5 mm,
and must be cut to the length required. They are frequently used however, for
bridging much larger spans such as those used to create internal partitions in
buildings (rolled steel joists or RSJs).

▼ The stiffness of beams ▲

The stiffness of a beam will vary depending on its exact shape. We looked at this briefly in Block 2 Part 1, when we considered the different stiffnesses experienced by a ruler depending on the way you flex it. You might like to look back at Figure 1.53 in Block 2 Part 1 to refresh your memory. Bending is harder when there is more material in the direction along which you are applying the bending force. This is why an I-beam (shown in Figure 2.22, so called because its cross-section is an I-shape!) is good at resisting bending.

Although making hollow beams is clearly a good idea in terms of weight saving, removing material will have some effect on the stiffness. The part of the beam where the stress is negligibly small is not a large volume, and this is the only bit that can be removed without serious consequences. In general removing stressed material from near the core means that the force that was being carried by it will have to be carried somewhere else instead.

The measure of a beam's stiffness arising from its shape is called the 'second moment of area', and is given the symbol I (but, confusingly, nothing to do with an I-beam: a beam will have a value of I irrespective of its shape). This quantity is different for different shapes of beam, as shown in Figure 2.23 below.

For any beam, the material from which it is made will also be important in determining the stiffness: remember that the stiffness is a combination of materials property and component geometry.

If a beam is loaded as shown in Figure 2.24, then there will be a deflection of the beam from its 'flat' position. The deflection as measured in the centre of the span is given the symbol Δ: the Greek capital letter 'delta' again, but this time it is being used on its own to represent deflection, rather than indicating a small change in a quantity. How much the beam deflects will depend on four things: the weight of what is placed on it, which gives a force F; the Young's modulus E of the material

that it is made from; the second moment of area I from its shape; and the distance L between the supports. The distance between the supports has a large effect – you may have noticed this if you've walked on a plank or board over a gap: slightly wider gaps appear to lead to dramatically increased sagging of the board under your weight.

Mathematically, for the simple loading shown in Figure 2.24 (known as 'three-point' loading, as there are two support points and one central load point), the deflection Δ is found to be equal to:

$$\Delta = \frac{F L^3}{48\, E\, I}$$

You can see the importance of the length separating the supports, as the deflection depends on the cube of this distance: so doubling the distance between the supports – which effectively means that the beam must be doubled in length – will increase the deflection eight times. Since the terms E and I are 'below the line' in the algebraic relationship, changing to a material with a *higher* Young's modulus, or a beam which is *stiffer by virtue of its shape*, with a larger I, will reduce the deflection.

Exercise 2.2

For the three beams shown in Figure 2.23, calculate the mass of a steel beam 1 m long, for W of 0.1 m and B of 0.05 m. The hollow beam and the I-beam have w of 90 mm and b of 40 mm. The density of steel is 7800 kg m^{-3}. Calculate the volume of the beam first.

What is the difference in deflection when such a beam is loaded? Investigate this in SAQ 2.5. Note that the I-beam and the hollow box section beam have the same value of I and volume. I-beams tend to be used in preference to hollow box sections because they are easier to form; there are also no internal areas where corrosion can start unseen.

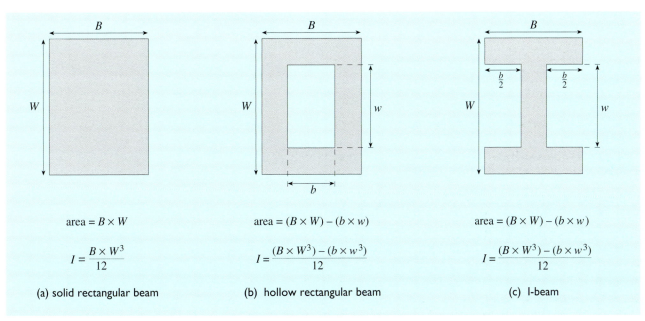

area $= B \times W$

$$I = \frac{B \times W^3}{12}$$

(a) solid rectangular beam

area $= (B \times W) - (b \times w)$

$$I = \frac{(B \times W^3) - (b \times w^3)}{12}$$

(b) hollow rectangular beam

area $= (B \times W) - (b \times w)$

$$I = \frac{(B \times W^3) - (b \times w^3)}{12}$$

(c) I-beam

Figure 2.23 Area of cross-section and I for different shapes of beam

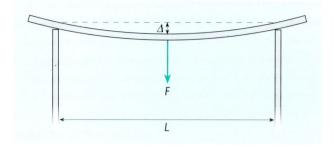

Figure 2.24 Simple 'three-point' loading of a beam

SAQ 2.5 (Learning outcome 2.8)

(a) Calculate I for the solid square section and the hollow box section beams described above, using the formulae in Figure 2.23.

(b) What will be the deflection of two beams, 1 m long between their end supports, when loaded as in Figure 2.24 with a force of 1000 N? The Young's modulus of steel is 210×10^9 N m^{-2}.

2.4 The patent document

Having discussed the background to lintels, I will now turn to the patent mentioned earlier: the Catnic patent. This example allows us to consider some simple structural concepts, but it is also useful in our story of inventions and patents, because this particular patent was challenged in court, and the ruling has had a major impact on the subsequent interpretation of patents in law.

The relevant patent can be found on the CD-ROM, and you should look at it to follow the discussion.

The patent is divided into several sections:

1   A description of the product;

2   Claims for the product concept;

3   Figures referred to in the description.

The wording of this patent is fairly accessible, though there is extensive technical detail, as one might expect. The language is rather convoluted, but as the concept is reasonably straightforward, and the detail is concerned with describing the form of the lintel which has been invented, it is not too difficult to read. The introduction is also nicely archaic! As for reading a standard, the detail contained within the patent is important.

Exercise 2.3

Read through the introduction to the patent (the first 25 lines of the first page), then answer the following questions. Your answers should not be too long: you are only looking for one or two relevant sentences in each case.

(a) What 'prior art' (i.e., the previous common practice) in this field of building technology is mentioned?

(b) What is the main problem indicated for the prior art lintels?

(c) What are the main aims of the lintel invention?

The patent requires detailed and accurate description of the nature of the invention. The lintel described is defined by a set of plates which are spot welded to one another, and they comprise

a first horizontal plate,	a first rigid inclined support member, and
a second horizontal plate,	a second rigid support member.

Figure 2.25 The Catnic Classic lintel. Here two of the plates are perforated and the internal cavity has an insulating foam block inside

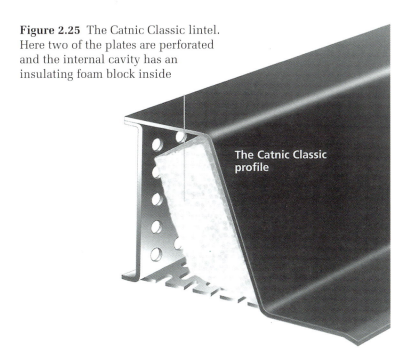

The plates are seen in the lintel in Figure 2.25.

We have already discussed in general terms what a lintel is and how it works. The patent goes into much more detail, though. The lintel is stated to be designed to fit an external wall of a building, where the wall consists of two vertical structures, or skins, separated by a cavity (as shown in Figure 2.26). Note that there is no reference to, say, 'brickwork', or 'concrete blocks' when describing the wall. One of the key elements in producing a patent is not to be too restrictive in terms of language. A patent that described a 'lintel for apertures in brick walls' could exclude the use of the patent to protect products used for steel walls, for example.

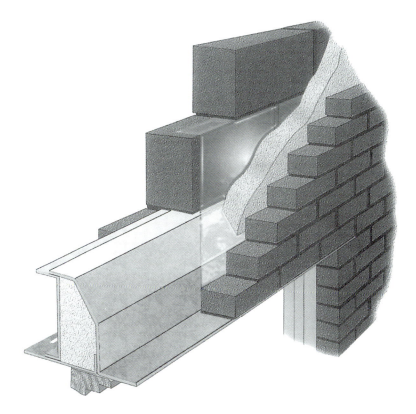

Figure 2.26 A heavy-duty Catnic lintel across a window in a cavity wall

The lintel spans the cavity immediately above the aperture which it bridges. The first horizontal plate supports the brickwork above it, while the second horizontal plate spans the cavity wall *and* the aperture. The apertures are created for doors and windows, or another fitting: again, this is not specified to ensure generality. The first rigid inclined support member faces in towards the cavity, and connects the two skins or walls. The second rigid support member faces the inside of the building, and is described and shown as 'extending vertically'.

The patent refers to a specific example of a lintel which is described in detail with reference to three figures at the back of the document. Compare these figures to the Catnic Classic profile in Figure 2.25. This specific example is known as the 'preferred embodiment' of the invention (on page 2 of the patent, from lines 47 onwards). Read this description of the preferred embodiment and then answer Exercise 2.4.

Exercise 2.4

(a) What is the preferred material of construction for the lintel?

(b) To what standard should that material conform?

(c) What other materials could be used in the lintel?

The preferred embodiment described in great detail in the patent should not delude you into thinking that this particular form of product is the only invention protected by patent. The claim which follows the preferred embodiment description is worded much more generally, and it is this claim which describes the inventive concept being patented; so the language is used to describe a more general idea of the invention.

Claim 1 is the most important claim in any patent document because it tries to configure the most general definition of the invention in an unambiguous and definitive way. This claim will attempt to define the concept in terms of 'essential integers': the essential features of the invention. There will probably also be subsidiary claims, which restrict the scope of Claim 1 by focusing on narrower and thus more specific features of the invention. The words used to describe particular parts are also chosen very carefully so as to convey very general meaning.

Claim 1 is challenging to read, but can be subdivided fairly simply to illustrate what the inventor is claiming. Subdividing it as follows makes this clear, with each of the essential points (or *integers*) given a separate paragraph.

WHAT I CLAIM IS:

1. A lintel for use over apertures in cavity walls having an inner and outer skin comprising

a first horizontal plate or part adapted to support a course or plurality of superimposed units forming part of the inner skin and

a second horizontal plate or part substantially parallel to the first and spaced therefrom in a downward vertical direction and adapted to span the cavity in the cavity wall and be supported at least at each end thereof upon courses forming parts of the outer and inner skins respectively of the cavity wall adjacent an aperture, and

a first rigid inclined support member extending downwardly and forwardly from or near the front edge adjacent the cavity of the first horizontal plate or part and forming with the second plate or part at an intermediate position which lies between the front and rear edge of the second plate or part *and* adapted to extend across the cavity, and

a second rigid support member extending vertically from or from near the rear edge of the first horizontal plate or part to join with the second plate or part adjacent its rear edge.

There are thus just four essential integers, each describing a side of a trapezium: this is what the cross-section of the lintel looks like, and the cross-section is constant along the length of the lintel. Each is specified as a 'plate or part or member', which from the description of the product means a level planar sheet of an unspecified material and of unspecified thickness.

Referring back to Figure 2.25, you should be able to see what each of these features refers to. The 'first horizontal plate or part' is the top bit of the lintel, the 'second horizontal plate' is the bottom, that sits across the brickwork, and the two 'rigid support members' at the sides that complete the structure and provide the necessary stiffness.

None of the terms used is difficult to interpret, and the meaning is clear from the description. Thus 'building units' is a very wide term which includes bricks, blocks and presumably even stone. The cavity is simply the gap between the two walls of brick or other building units.

The claims that follow Claim 1 are intended to give specific definitions of a product that conforms to Claim 1. The reason that such definitions are not included in Claim 1 is that they would then restrict the patent to the specific case.

The other claims narrow the breadth of Claim 1 in very specific ways. Thus Claim 3 specifies that the plates may be integral with one another, that is at least two plates may be simply bent to shape from a larger sheet to form a single unit. Claim 4 extends the concept to include a 'lazy zed' configuration. The claims become narrower and narrower and end with the preferred embodiment and a cavity wall within which it sits. Note that each claim starts with a previous claim, so the degree of narrowing is well defined.

The reason why patents are important is that they provide the inventor with a monopoly for 20 years from the date of application. So, in the case of a patent granted in the UK, if anyone within the UK copies the idea, they can be sued for patent infringement.

For a successful new invention, the profit can be substantial, so that others may be tempted to design around the patent. Since a patent is a legal document which defines the monopoly granted by the State, it is vital that the boundaries of that monopoly are defined very precisely. This is why great effort is extended in trying to value the widest claim – Claim 1 – as clearly and as wide as possible. The danger for the patentee is that if the claim is too wide, it can include an earlier concept or product. If this happens, the patent is null and void because it is 'anticipated' by the earlier idea or product. We will meet just such an example later.

2.5 The Catnic patent challenged

Shortly after the original Catnic lintel was developed and patented, a challenge to the monopoly provided by the State came from Hill and Smith, a company who had hitherto made crash barriers for motorways. With a prospective reduction in the motorway construction programme, the company was looking for alternative products to construct from sheet steel. As is common practice in most manufacturing industries, the starting point was a search of competitor's brochures. One product, that of the Catnic lintel, looked interesting, because the product was of very simple shape, and it would not need a great deal of tooling to convert sheet steel into a lintel.

The problem came with how to avoid infringing the claims of Catnic's patent. Clearly, if Hill and Smith copied exactly the form described in words in Claim 1 of the patent, it could be sued and if it lost, would have to pay substantial sums of money in royalty fees to Catnic. Since Claim 1 of the patent described a *vertical* plate for the rear support, the company reasoned that a plate *slightly off the vertical axis* would avoid the requirement of the claim. The claims of a patent are critical because they define the limits and scope of an invention. Remember too, that claims are written in the English language, and so there may be some ambiguity in the terms used. However, the person who writes the claims (normally a patent agent), tries to draft them as tightly as possible so that any technical reader will be clear what the monopoly covers.

Claim 1 of the Catnic patent is unfortunate in apparently limiting the claim to a rear vertical plate. So if one wanted to avoid infringement, it would be easy to design the back plate to be slightly off-vertical (Figure 2.27). This is what Hill and Smith actually did, and the resultant court case became a cause célèbre for patent law in England. The outcome of the case is still cited today as a legal precedent for judgement in cases of patent infringement.

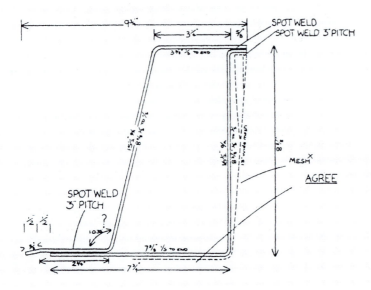

Figure 2.27 The Hill & Smith lintel that was alleged to infringe the Catnic patent

Hill and Smith lost in the patents court, won in the Court of Appeal, and the case proceeded all the way to the House of Lords before the issue of interpretation of the words of the patent was finally determined. Eventually, the rather elementary issue of the meaning of the term 'vertical' was decided by the House of Lords in a judgement which has affected the interpretation of all subsequent patents. Their reasoning was based on sound engineering principles. They said that inclining one support from the vertical by a few degrees was immaterial to the function of the device. If the maximum imposed load gives a force F acting vertically downwards, then a lintel inclined at 6° from the vertical can bear a force of $F \times \cos 6° = 0.9945\ F$ (see **▼Forces and angles▲**). The change in the load capability for the Hill and Smith lintel design is therefore trivial: only 0.55%.

The Law Lords further went on to say that patent claims should be interpreted purposively: that is the words should be read for the intent of the patentee, and any slight or immaterial changes to the design should fall within its ambit. All patent claims are now read with this judgement in mind. Of course, with the benefit of hindsight, using the phrase 'approximately vertical' would probably have achieved the original objective of the patentee without the need to go to the House of Lords!

▼Forces and angles▲

In the case of a box-shaped lintel, the vertical (or near vertical) plates must bear the downward force on the top horizontal plate. Inclining one of the plates at a slight angle increases the forces within that plate for a given load on the lintel by a geometrical effect. The design is restricted by the size of forces in any component, so the design limits will be reached at a *lower* load if a vertical plate is slightly angled – but can we say by how much?

Here's an illustration based on a more familiar example: that of drawing a curtain across a window. In this example the 'useful' force is in the *horizontal* direction. Any vertical force just serves to pull the curtain taut without moving it sideways. For simplicity let's think about the sort of curtain that is hung from rings that are free to slide along a pole. To draw the curtain you have to provide a horizontal force just sufficient to overcome any friction between the curtain rings and the pole.

Often the top of a curtain is out of reach or otherwise inaccessible. Like me, you've probably still managed to draw the curtain by pulling at the curtain from lower down; see Figure 2.28(a). How does pulling largely downwards on the curtain provide the necessary horizontal force?

Well, any single force can be broken down (or *resolved*) into two others which, acting in concert, produce the same effect. The usual and useful way to do this is to choose one force in the direction you are most interested in – here that would be a force directed horizontally along the curtain pole. Then the second force is chosen to be perpendicular to the first, in this case vertically down.

We can find the relative size of the component forces by means of the following simple rule: the amount of a force F in some direction at an angle θ (the Greek letter theta is commonly used to represent angles) to it is:

$$F_{\text{at angle } \theta} = F \cos \theta$$

If $\cos \theta$ (pronounced 'coz theta') is a familiar expression then skip forwards to the next paragraph. Cos θ is a mathematical function more properly called the cosine function. What the function does is to provide a number

between plus and minus one whose value is uniquely determined by the angle theta; here theta is measured in degrees. Effectively the function is a sort of look-up table. If theta is zero or 360° the cosine of theta is plus one. If theta is 180° the cosine of theta is minus one. Other angles give a range of values that smoothly pass between the limits of plus and minus one as theta is varied. The graph in Figure 2.28(c) shows the full picture. You should have a calculator with this function on it. Check a couple of values now, using your calculator to find the cosine of 60° and of 45° and compare the results with Figure 2.28(c).

Let's put this resolved force idea to work. In Figure 2.28(b) I've shown the force that must be applied to the curtain to draw it is F_a and I've resolved it into two components. It is the component of the force along the pole that will do the actual moving of the curtain. The other component is vertically downwards and acts as a load on the pole fixings, adding to the weight of the curtain. Looking at the horizontal component, our rule for resolved forces says:

$$F_h = F_a \cos \theta$$

Look again at Figure 2.28(a) and 2.28(b).

Exercise 2.5

Since we've chosen the angle between the component forces to be θ the angle between the applied force F_a and the vertical is $(90 - \theta)°$. What is the corresponding expression for F_v the vertical component of F_a?

We're now in a position to see how applying F_a at some angle θ to the horizontal gets the curtains moving. For any given angle θ we need F_a large enough to make F_h, the horizontal component, equal to F_d, the sliding friction. So

$$F_h = F_d$$

or

$$F_a \cos \theta = F_d$$

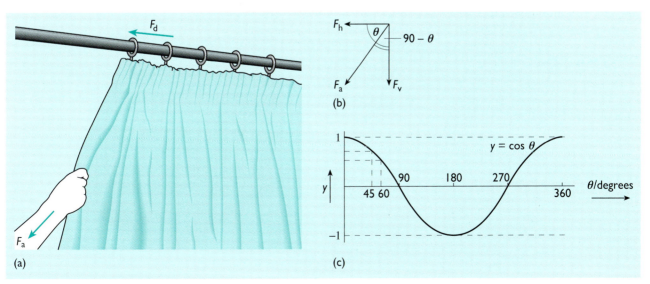

Figure 2.28(a), (b), (c) Drawing forces

Rearranging this expression we can deduce that the amount we apply to the curtain is:

$$F_a = \frac{F_d}{\cos\theta}$$

That feels right. If θ is zero the curtain is pulled directly by all of the applied force:

$$\cos(0) = 1$$

so

$$F_a = F_d$$

If θ is say 60°, the curtain is only pulled by the horizontal component $F_a \cos(60°)$; since $\cos(60°) = 0.5$ the applied force has to be *twice* that needed to match to sliding friction $F_a = F_d/0.5 = 2F_d$.

Intuition and experience are also satisfied by the conclusion that pulling vertically down will never get the curtain to slide: $\cos(90°) = 0$ so the horizontal component is always zero, no matter how hard you pull so it can never supply F_d.

Curtain material will tear if subjected to a sufficiently large tensile force like F_a in Figure 2.28(a). The proper engineering design of curtains ought to stipulate a maximum safe angle of pulling. (Of course the wall fixings may fail first.)

Returning to the lintel I hope you are now at least not surprised that inclining a vertical plate at an angle θ means for a given load F_L the necessary force in the plate (F_p) is

$$F_p = \frac{F_L}{\cos\theta}$$

A design specification that limits forces in the plate to F_{pmax} now translates into a safe working load

$$F_{Lmax} = F_{pmax} \cos\theta$$

And that is where we started. The load bearing capacity is reduced by a factor of $\cos\theta$.

2.6 Catnic lintels

The success of this product in achieving its design goal reflects that such hollow structures had long been widely used elsewhere in engineering structural applications. The idea has also been applied with great success to modern road bridges, where the carriageway is built of flat hollow sections bolted together to form a continuous structure (Figure 2.29). A famous example from the Victorian era is the Britannia Bridge crossing the Menai Straits, made by enclosing the rail track in a steel box section (Figure 2.30). So if the general idea is not new, what is inventive about the Catnic lintel?

Figure 2.29 The carriageway of the Humber Bridge was constructed from hollow sections

Figure 2.30 The Britannia Bridge

Inventions are about new devices to solve problems in very specific fields: building technology in this case. The *form* Catnic chose was novel, with nothing shaped just like the lintel they developed (Figure 2.21). For example, an earlier sheet steel lintel made by the company Dorman Long lacked the stabilizing effect of the lower base plate. The second step in proving invention is establishing that the concept is new and original as applied to this narrow area of building technology. Although patents are often challenged by infringers, it is rare that challenges have to be decided by court action. However, this is precisely what happened with the Catnic lintel. Because Catnic won the case and its patent was found to be valid, it became the market leader, developing the product in many different ways. This is true of all fundamental advances in technology: once a new concept has been formulated, the way is open to refining the concept to suit a variety of different applications.

2.6.1 Lintel sizes

So what were the first developments from the basic design? Making a range of standard fittings to suit different window widths (but at a constant depth of two bricks plus standard cavity) is a way in which choice can be offered to the builder (Figure 2.31). The thickness of the sheet steel can also be varied from 1.6 mm up to 4.0 mm, depending on the load supported by the lintel. The thickest lintel gives a safe working load of 6.9 tonnes, when uniformly distributed over the top surface of the lintel.

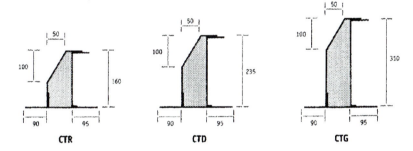

Figure 2.31 A range of lintels

2.6.2 Lintel protection

One problem with all steel structures is the potential for corrosive attack from the atmosphere, especially rusting by contact with water. So what is the solution? There are several different ways of protecting exposed steel, some of which you may be familiar with if you are a car owner. They include:

    galvanizing (plating with a zinc layer)

    painting

    polymer powder coating

and Catnic actually uses a duplex (two-layer) protective system, based on a thermosetting polyester powder coating over a galvanized inner coating to the steel.

2.6.3 Cold bridging

The possibility of water penetrating into buildings is a problem which the cavity in the wall is designed to overcome. But of course, providing a cavity also weakens the structural integrity of the wall. Bricklayers use wall ties placed at regular intervals in the structure as it is built to overcome this. They are usually simply short lengths of galvanized steel (or stainless steel) provided with spread ends and a twist in the centre. The function of the ends is to provide a secure anchor, while the twist helps stop liquid (water) bleeding across the tie. This is the problem of 'cold bridging'. So if there could be a problem with small wall ties, could the same problem affect large steel lintels with a continuous plate bridging the cavity?

Several design features can be incorporated to minimize or eliminate the problem. First, the lower inner part can be slotted to minimize the moisture path (Figure 2.25). Secondly, the plastic coating effectively means that each lintel is a damp-proof layer, so rising damp is prevented from rising above the lintel itself. Finally, the outer leaf is provided with a slanting edge, so that any moisture will drain away back to the exterior of the building.

2.6.4 Other designs

If a box section can give strong support to superimposed loads, are there any other designs which would play the same role? Another profile made by Catnic, known as the 'Cougar', which gives similar support is even simpler in design (Figure 2.32). The base of the profile is open, so that the lintel can be made in a single manufacturing operation, only involving pressing sheet steel. Wide leaves are left at both sides so that a good joint is formed with the surrounding brick work. Since there are no joints in the lintel itself, manufacturing is easier, and the path for ingress of water is much more convoluted. One problem the open end presents is that there is no material against which to plaster. So the space is filled with polystyrene foam which is corrugated to provide a key to plaster (should it be needed). The polystyrene is highly insulating, so reducing heat loss through the lintel.

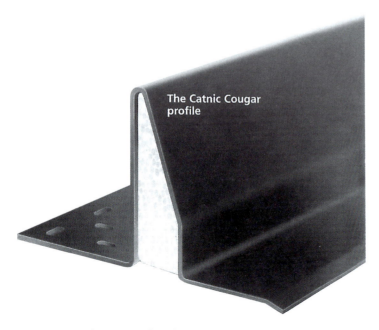

Figure 2.32 The Cougar lintel

SAQ 2.6 (Learning outcome 2.7)

Does the Cougar lintel just described and shown in Figure 2.32, fall within the scope of Claim 1 of the original Catnic patent of 1968? Use the earlier discussions to help you decide whether or not the Cougar lintel possesses all of the essential integers specified in Claim 1, giving full reasons for your answer. Would the addition of any further parts to the Cougar design bring it within the boundaries of Claim 1 (or any other of the claims) of the earlier Catnic patent?

2.7 Summary

In this section we have looked at an example of a structural product – a lintel – and seen how a company developed the concept and embodied its idea within a patent. This particular patent has additional interest because it was rigorously examined in the law courts, and the judgement associated with it had important implications for the way patents are interpreted when they are challenged.

The inventive step was the use of sheet steel to make a hollow beam for the lintel, which was an attractive replacement for solid lintels. The design in this case has been developed by the company, to the stage where the latest designs are far from that which was specified in the original patent.

Such incremental design is usual in product development. We will look at another patent-related example of this in the next section.

3 Patenting a consumer product

3.1 Introduction

In this section I will look at another aspect of the design process: that of
incremental design. The example I will use is that of the lawnmower, a
product that has been in existence for quite some time, and which has
evolved gradually into the machines that we know today. You will probably
have noticed a change in at least the form of the basic lawnmower during the
last ten to twenty years. Such changes are difficult to classify as inventive,
unless there is some clear innovation in the structure or operation of the
product.

I will illustrate this by looking at the outer shell of the mower, but first let us
look generally at the mechanics of such a shell, and the general function of
the mower.

3.2 Monocoque products

The use of thin sheets of material to bear imposed loads is a widespread
construction method used in many products, and is not restricted to building
or industrial products. The method is known as *monocoque* construction. It
became widely used in car-body construction in the early days of motoring.
The key change occurred in the 1920s in France, but was adopted almost
universally for the then-new mass manufacturing methods for cars. Instead
of a heavy and rigid base (the chassis), to which the engine and body were
attached, the monocoque comprised all of the body of the car, so the loads on
the car whilst driving were borne by the whole body, including the roof, and
not just by the base unit. The difference between the two approaches is
shown schematically in Figure 2.33.

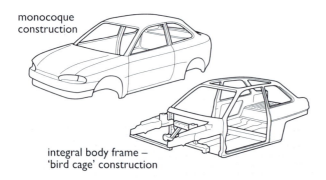

Figure 2.33 Monocoque and space frame construction in
cars. The monocoque uses the 'outer body' of the car for
its stiffness. The space frame uses a central chassis and
rigid beams, onto which the exterior panels are hung

Similar changes occurred in aircraft construction, where low weight was
clearly a main design objective. In the first days of the Wright brothers,
aircraft were built as a *space frame*, where the loads were borne by struts and
ties. The main body and wings were then covered with textile fabric doped
with lacquer to make them airtight. But all the loads imposed by flying were
supported by the struts, the thin skin taking very little.

The first monocoque aircraft used thin aluminium sheet (often corrugated to
provide extra stiffness) to form the skin when attached to a light alloy frame,

and the skin bears a substantial proportion of the flying loads. The method gained status during the Second World War, when the wings, fuselage and tail of the Spitfire fighter were built in monocoque form.

The application of the monocoque structure has expanded substantially with the growth in products designed for use by consumers in a domestic environment. Such products range from furniture and powered cleaning products to garden mowers.

The market for domestic consumer products is one of the most active for inventors, for a variety of reasons. They include:

1   the sheer size of the potential market (millions of households around the world);

2   demand from users for innovation and improvement;

3   the speed with which a basic concept can be turned into a manufactured product.

One such market is gardening, virtually a leisure activity for much of the UK population, and one particularly favoured owing to the mild climate of the UK. We will look generally at one item of gardening equipment – the lawnmower – before looking in detail at the design of the structure, and a case involving alleged infringement of a patent for the structure of a lawnmower.

3.3 Lawnmowers

The lawn is a central feature in most gardens, both for reasons of tradition (in the UK) and because of the fecundity of grass growth. So what are the principal garden tools? Activities such as digging, hoeing, trimming, and most of all, mowing, have been mechanized to an increasing extent. In past times, they may have been performed by hand or using animal power, but today, labour is not cheap, and mechanization offers faster time-to-completion, greater consistency of the final result and a reduction in effort.

> **SAQ 2.7** (Learning outcomes 2.2, 2.3, and 2.4)
>
> Two types of machine used in lawn maintenance are the mower (which can be obtained in different designs, Figure 2.34 on page 74) and the edge trimmer (or 'strimmer').
>
> (a) Describe the function of these machines, indicating their main function, and how they achieve that function (from your own experience, or observation).
>
> (b) What problems do the inventors of such devices wish to solve? (This is different from just 'cutting the grass': try to envisage how lawns were mowed before each of the designs was formulated, and what was innovative about each design.)

3.3.1 The hovermower

Much innovation in mechanized garden products has occurred only since about 1970, and many products now utilize plastic casings or enclosures. The reasons are not difficult to appreciate. They are generally free from corrosion, are lighter than metal alloys and, if well-designed, can present a pleasing appearance (depending on your taste). Their widespread introduction dates from the invention of the hovermower, an invention based upon the principle of the hovercraft, invented by Cockerell in the mid-1950s.

The basic principle at work in hovering devices is based on the lift provided by an air current. The lifting force generated is proportional to the square of

Figure 2.34 Different types of lawn mower (cylinder mower, hovermower, wheeled rotary mower)

the air velocity, and a high velocity air current is needed to drive such devices (see ▼**Linear and power-law relationships**▲). It can be provided by a small electric motor driving a fan blade, which must be directed down onto the ground. The essential principle of the hovermower is very simple, and uses the same motor which gives the air blast to power the rotating grass-cutting blade. The two blades are mounted coaxially (i.e., they rotate on the same axis), and the air blast is directed under a hood which covers the cutting blade (Figure 2.35). In use, the blast lifts the hood from the ground, so the whole device can be moved in any desired direction. As it moves over the lawn, the blade cuts the grass.

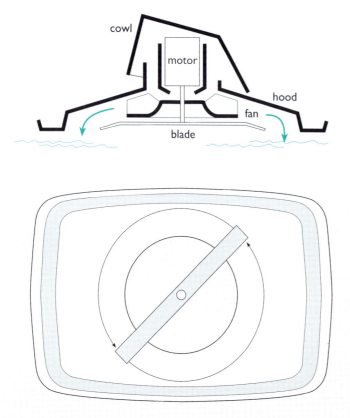

Figure 2.35
The hovermower

▼Linear and power-law relationships▲

The relationship between physical quantities is not always straightforward and *linear*. A linear relationship is one where two quantities scale together. So, for example, doubling the length of a wall will require double the number of bricks (assuming the height and width of the wall is unchanged).

If the bricks are used to pave a square patio, though, doubling the length of the patio side means that *four times* as many bricks are needed, as the area being paved has increased by that factor (Figure 2.36).

This is known as a power-law relationship. In this case the number of bricks is related to the square of the length of one side of the patio; it is related to l^2 (i.e., $l \times l$).

More help on power-law relationships can be found in the *Sciences Good Study Guide*, p. 353.

In the case of the hovermower, the lift force is proportional to (i.e. goes up in the same proportion as) the square of the air velocity. So this means that if the mower is twice as heavy, an increase in air velocity of just over 40% is needed to generate the required lifting force (the square root of 2 is 1.414: so an extra 40% gives $1.4 \times 1.4 \approx 2$ times the lifting force). This will require a more powerful, and hence heavier, motor; so there is a big incentive to keep the mower as lightweight as possible – hence the use of plastics for the body.

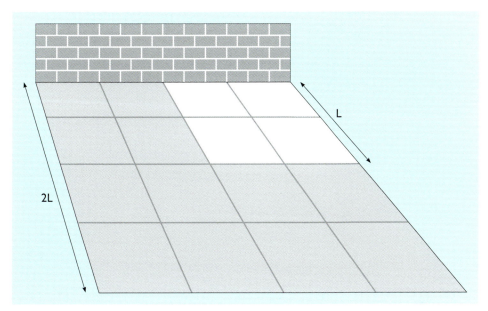

Figure 2.36 Paving a square patio

It is the low cost and ease of use which led to the great popularity of the hovermower in the mid-1970s, when it became widely available to the public. The hood is a key component of the machine because it protects the user from the blade and from flying objects thrown up by the blade during mowing (stones are a particular problem). The mower can move over steep slopes and rough ground where no other mower will work, a unique attraction. They are also better able to cut wet grass, which defeats cylinder mowers.

SAQ 2.8 (Learning outcome 2.4)

(a) List the main functions of the hood on a hovermower when used to mow grass. Use Figure 2.35 to help you think about the duties a hood performs in service. What other functions could be considered important?

(b) What materials property requirements are needed in the hood to achieve each of those functions?

There are several possible material solutions to the design of the hood, and the first prototypes indeed used thin sheet steel, as well as GRP (glass-reinforced plastic), a material based on polyester thermosetting resin (a plastic) reinforced with glass fibre. Thin sheet steel rusts easily despite protection with paint or galvanizing, mainly because stones thrown up by the blade will quickly penetrate the protective layer. Composite materials – like fibreglass, where reinforcing glass fibres are mixed into a polymer matrix – are feasible, but are an expensive option. Thermoplastics are the best solution, because they can be mass produced quickly using a process route called injection moulding, and are tough and stiff enough to withstand imposed loads. The material actually used in most hover mowers is a polymer called ABS (acrylonitrile-butadiene-styrene, after the three different types of compounds used in the polymer chain structure).

Since the first introduction of the hovermower in the late 1960s, dating from a patent first filed on 1 October 1964, the idea has been developed further with grass-collecting hovermowers (Figure 2.37). One problem with the original design was that the cut grass was left on the lawn, a problem the new version neatly solves by using part of the air current generated to suck the grass into a container above the motor.

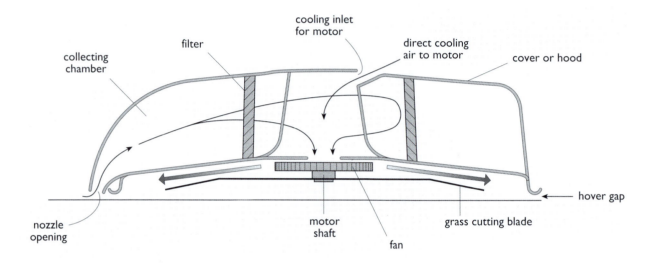

Figure 2.37 Grass-collecting hovermower

3.3.2 Wheeled rotary lawnmowers

However, the quality of finish to the lawn is limited using a hovermower: for one thing, there is no control over the height to which the grass is cut. A wheeled rotary mower does give control, and heavier, more powerful motors (such as petrol engines) can be used to give a wider cutting area beneath the blades. Grass collection is also easy with a rear collection basket. Overall product weight is now less critical, and, until relatively recently, cast aluminium bodies were used widely. They are rather heavy, and many manufacturers now use plastic instead. There is another, more subtle reason why plastics are preferred. Much more detail can be incorporated into the hood, so design freedom is much greater.

3.4 Monocoque mowers

For a hovermower, we have seen that a motor needs to be mounted in a casing of some type. A designer has essentially three options for the structure, as shown in Figure 2.38: a monocoque, a platform and a space frame. The shells bear the load in the first two options, while the struts bear the load in the space frame.

Until about the late 1980s, design of mowers was rather restricted by numerous patents which had been taken out on various different design concepts for the body. One large manufacturer, Flymo, introduced two new

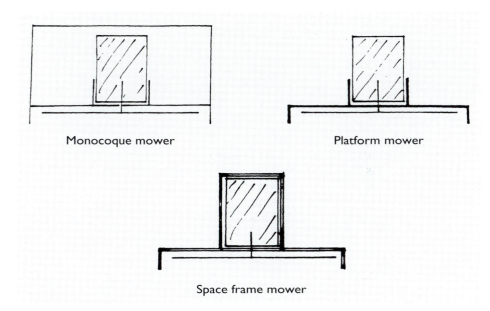

Monocoque mower   Platform mower

Space frame mower

Figure 2.38 Different structures for the casing of a mower

designs in plastic and was sued by a competitor (Black & Decker) for infringement of one of its patents.

However, when the case came to trial, the problem of using a patent which was close to the prior art (i.e., that which is known already, either from earlier patents, or the published literature) became very clear. Remember that in the list presented earlier for assessing an invention, an important factor is that the invention must be non-obvious. If a product already exists which falls within the claims of the patent, then the claims of the patent are not inventive. The problem is that of interpretation of the all-important Claim 1 of the patent. In order to catch an infringer, the patentee may have to widen the interpretation of the claim: as in the case of Catnic, where 'vertical' was eventually taken to mean 'substantially vertical'. But if the patentee does widen the claim, there is a risk that it may then include the prior art. If it does, then the patent must be void because all patents must be new and original.

The classic defence to an allegation of patent infringement is to show that each step taken in the design of the alleged infringing product is trivial and non-inventive, and is known as ▼**The Gillette defence**▲ (on page 78). However, in Flymo's case, another defence was used: that a known prior art patent was infringed if the Flymo product was in infringement. If that were true, then the Black & Decker patent would be void.

3.4.1 The Rasenkatz patent

The patent Black & Decker used to sue Flymo in 1990 is provided on the CD-ROM. The patent has a priority date of 16 June 1977, when it was filed in Germany, and we will refer to it as the Rasenkatz patent.

In order to use thermoplastics successfully, the inventor of the Black & Decker 'Rasenkatz' (German: literally 'lawn cat'!) mower encased the motor between two shells which, when joined together, formed a monocoque much stiffer than either of the two subunits. This principle can easily be demonstrated by considering a sandwich box which is sealed by a plastic lid. Deformation of the open plastic box is much easier than when closed with the lid.

A specific embodiment of the claim is shown in Figure 2.41, taken from the equivalent US patent. Large manufacturers frequently patent their products worldwide, but especially in the USA, Japan and Europe; this provides them with protection of a monopoly in the most important economies of the world.

▼The Gillette defence▲

A frequent defence by alleged infringers of patents is what has become known as a 'Gillette defence', after a case in the 1900s when the manufacturer Gillette of the US sued Butler, an English company, for infringing its patent involving safety razors. The case came to trial, was appealed, and then further apppealed to the House of Lords. It has become embedded in English patent law as a good defence for a product against a patent infringement action.

The defendant has to show that the design was arrived at by a series of steps, each of which was obvious at the time it was made, and therefore non-inventive. Clearly then, the final design must be non-inventive, and cannot infringe any patent in force at the time. (The patent which is used to sue the alleged infringer may then seem to be non-inventive also, but this argument can fail depending on the interpretation of the patent by the court. In the Gillette case, the safety razor patent was held to be valid, but the Butler product did not infringe.)

The Gillette razor is shown in Figure 2.39, together with the Butler product which was alleged to infringe the Gillette patent (Figure 2.40).

The key difference between the two products lay in the shape of the blade when mounted to be used for shaving the face. In the Gillette, the blade is bent by the metal cover plate, while the Butler razor holds a thicker blade in a flat position. The House of Lords held that each feature of Butler was either obvious, or common general knowledge and non-inventive. It could not infringe Gillette's patent, because the blade was flat and not curved. This decision set the doctrine as a classic piece of case law, which is still very relevant and still cited today.

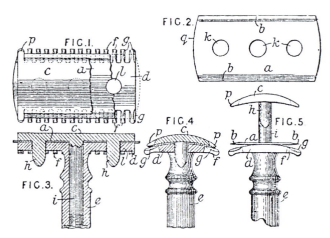

Figure 2.39 The Gillette razor

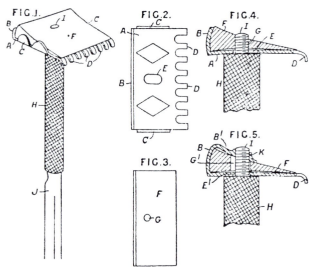

Figure 2.40 The Butler razor

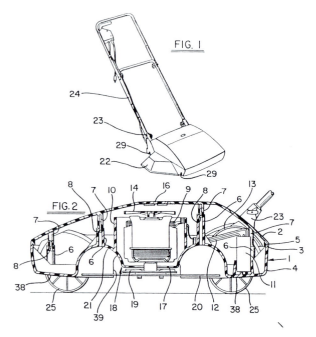

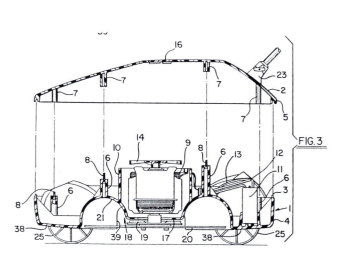

Figure 2.41 Rasenkatz lawn mower

Have a look at the Rasenkatz patent document. Concentrate on the preamble on the first page, and the actual claims starting at the bottom of the second page. You can skip the section on page 2 referring to the detailed description of the diagrams.

SAQ 2.9 (Learning outcomes 2.1, 2.3, and 2.6)

Read the introductory preamble on the first page of the Rasenkatz patent document and then answer the following questions.

(a) What prior art is mentioned in the preamble (before the claim)?

(b) What does the patent document say are the main problems with the prior art mowers?

The Rasenkatz mower tackles the perceived cleaning problems with earlier mowers by providing a smooth and relatively flat lower edge (lines 34–37, page 3), and smooth, easy-to-clean surfaces especially in the cutting region (lines 64–66, page 3). The problems with the wheels are solved by providing pockets or chambers in the lower housing within which the four wheels are recessed (lines 107–115, page 3).

Claim 1 of the patent is as follows. As before, I have broken it down into sections to highlight the different parts.

WHAT WE CLAIM IS

1   A rotary lawnmower comprising:

(i)   a dish-like body defining the chassis of the lawnmower and having a substantially continuous bottom wall, at least a portion of which is substantially in the plane of a cutting blade of the lawnmower, and a wall extending upwardly from the bottom wall to define the outermost peripheral sidewall of the body;

(ii)   a motor equipped with an output drive shaft having a free end, said motor being mounted in said body so as to cause said free end to penetrate said bottom wall for carrying the cutting blade thereon; and,

(iii) a cover coextensive with the uppermost edge of said outermost peripheral side wall and engageable therewith to define, conjointly with the dish-like body, a housing enclosing said motor.

There are thus just three essential integers which describe the inventive concept, a lower housing (i), an upper housing (iii) and a motor fitted with a blade (ii).

The lower housing (chassis) is defined as a 'dish-like body', which is a good description of the chassis shown in Figures 1 and 2 of the patent. However, the figures show the preferred embodiment, so the definition in the claim must go wider to cover all chassis with dish-like bodies. The 'wall extending upwardly' simply means the rim around the lower housing which forms the edge of the lawnmower.

The motor defined in Claim 1 is unexceptional, the drive shaft connecting to the blade.

The cover (or hood) also seems straightforward and presumably can be of any shape as long as it:

(i)   is coextensive with (joins to) the outer peripheral sidewall of the chassis and

(ii) engages with it to define

(iii) the housing enclosing the motor.

What the inventor is saying with this patent is that it is the form of the lawnmower which is inventive. Clearly this is not claiming to be the first time that someone has invented a lawnmower; in the same way that Catnic were not claiming to be the first to invent the lintel.

The subsidiary claims narrow down the very general specification of the invention of Claim 1 in the same way as for the Catnic patent. They specify the shape of the cover (Claim 2) and the material of construction, now explicitly stated to be a 'thermoplastic synthetic material' (Claim 5). An interesting and apparently substantive specification appears at Claim 10, which restates Claim 1 but now includes

' a set of wheels'

in addition to the chassis, cover or hood and motor. It thus appears that Claim 1 does not include a set of wheels as being an essential integer of the invention. As before, the final claim is simply the preferred embodiment already described and shown in the five figures of the patent.

A specific embodiment of the claim is shown in Figure 2.41, and the shape of the body seems to be reasonably included by the claim. But the key term used to describe the shape is 'dish-like'. This does not seem like a technical term, so what exactly does it mean? A sensible interpretation of the term dish would be a saucer or shallow plate with an upturned edge at the periphery (as the patent demands).

Black & Decker claimed that Flymo had produced a mower design which infringed its Rasenkatz patent. The question of infringement is usually considered first during a patents trial (and the subsequent judgement). The problem is two-fold. First, how are the claims likely to be construed by an engineer (the 'skilled person')? He or she will interpret the claims by reading them in conjunction with the description (just as we have done). Then the infringing product is examined in detail to see if it takes all the essential integers of Claim 1. Since this is the widest claim, it is the most crucial for deciding the issue.

SAQ 2.10 (Learning outcome 2.7)

Figures 2.42 and 2.43 show schematic diagrams of two mowers: the Flymo RE 42, which was the mower alleged to have infringed the Rasenkatz patent, and a Zundapp mower from 1969.

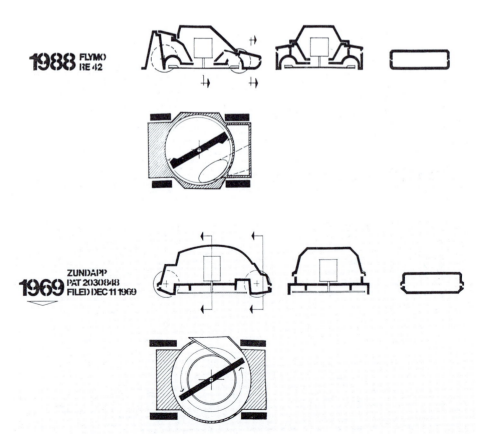

Figure 2.42 The Flymo RE 42 lawnmower

Figure 2.43 The Zundapp mower

Looking at each of these mower designs, try to match them against the chassis, motor, and hood features from Claim 1 of the Rasenkatz patent. Do they appear to infringe any part of Claim 1 of the patent?

At trial, it is the task of the claimant to show that the defendant's product falls within the patent claims. But in this case, if Black & Decker argued that the Flymo RE 42 fell within the meaning of Claim 1 of Rasenkatz, there was the distinct danger that its interpretation also included a much earlier patent, for the Zundapp mower. This is the patentee's dilemma: too wide a boundary to Claim 1 and it catches prior art and is invalid; but too narrow a boundary, and it fails to catch any possible infringers.

The Zundapp mower appeared to be caught by Claim 1 of Rasenkatz, and supporting evidence came when the evidence showed that the box and cover could also be injection moulded: the box of the Zundapp was originally cast aluminium, but could easily be replaced by mouldable thermoplastic. The credibility of the patent was therefore further undermined: it did not appear to be novel at the time of its issue.

3.5 Design development sequence

One way of defending an attack of infringement is to show that the basic concept of a patent is obvious at the priority date of the patent, and that hence it is obvious in the light of prior art, and invalid. This is normally the task of an independent expert witness, who can advise the court in its approach to interpretation or construction of the patent. The argument at the 1990 trial proceeded as follows.

The first step is to consider the design for function, i.e. cutting grass using a motor-driven blade. The motor and blade must be supported safely and rigidly, and there must be a partly open space at the base of the mower for the action of the blade on the grass.

I have already indicated that the mower uses a monocoque construction. For a device of this type, there are essentially three options for the structure, as shown in Figure 2.38: a monocoque, a platform and a space frame. The shells bear the load in the first two options, while the struts bear the loads in the space frame.

Once a monocoque has been settled on, the design is developed further to have a stiffening flange around the rim (Figure 2.44). However, if injection moulded, a single, closed monocoque is impossible to make by moulding since the method involves injection of hot polymer into a metal tool with a shaped cavity. The mould is usually in two halves which join together, and the shape of the product must allow for a mould that can be separated, and

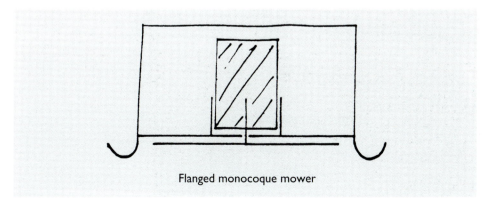

Flanged monocoque mower

Figure 2.44 Monocoque-style design of mower with a stiffening flange around the rim

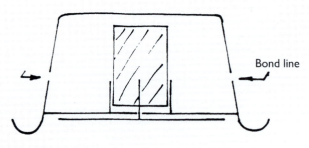

Thermoplastic flanged monocoque mower

Figure 2.45 Monocoque mower body made from two separate pieces which are joined together

the product removed from it. Shapes that have complex re-entrant angles cannot be made, as they are simply impossible to remove from a mould. For a monocoque mower body, then, it must be made in two or more separate pieces which are joined together subsequently, like that shown in Figure 2.45.

This development sequence was presented as being a stepwise, obvious approach, with no inventive steps.

3.6 Design development at Flymo

Figure 2.46 shows the actual development sequence at Flymo, from the initial hovermower to the two models alleged to infringe the Black & Decker patent. Since a Gillette defence was used by Flymo, it was important for the Claimant to question the Chief witness-of-fact for Flymo (its chief designer) about the sequence.

One interesting point which did emerge from cross-examination, however, was that the stiffness of the Flymo models derived not so much from the lid attached to cover the motor, but by the grass-collecting duct at the rear end of all the Flymo models (seen best on the lowest drawing in Figure 2.46). In fact

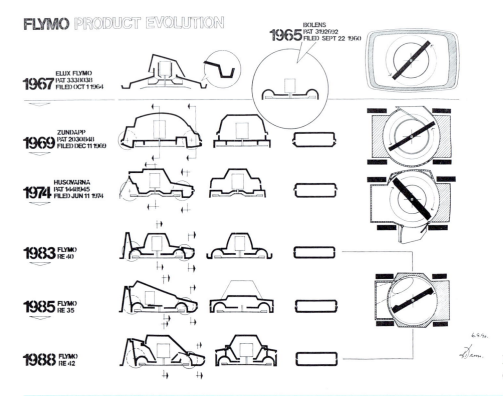

Figure 2.46 Product design evolution at Flymo

the mower operated perfectly well when the top was not attached at all! This demonstrated quite conclusively that the stiffness of the complete body derived as much from this duct as the attachment of the lid.

Each step that the designer took in developing the range of models was therefore demonstrated to be non-inventive and thus the Gillette defence succeeded at trial.

3.7 Judgement

Did the Flymo models actually infringe Claim 1 of the Black & Decker patent? It really depended on how the court interpreted the term 'dish-like', and what was suggested by the figure from the patent itself. In the end, the court took the view that the term 'dish-like' was indeed saucer-shaped and could not be extended to the wedge-shaped bodies of the Flymo models. The Flymo models therefore did not infringe the patent.

But was it a valid patent? We have seen above that the patent had been challenged by the Flymo expert as being obvious, in that a qualified and experienced designer following a simple design sequence using well-known principles would arrive very easily at the patent mower. This was not enough in itself to destroy the patent, because, for example, the expert admitted that he was not a lawnmower designer (he was an OU academic in fact!). However, the similarity of the structure of the Zundapp to the expert's preferred design was very clear, and this reinforced the conclusion that the patent was indeed obvious.

A second point which arose during the trial concerned the strict requirements of Claim 1: nowhere in the claim is the need for wheels mentioned.

SAQ 2.11 (Learning outcome 2.7)

Have another look at the hovermower shown in Figure 2.35.

Looking at the hood, motor, and cowl of the mower, decide whether, in your opinion, whether it infringes Claim 1 of the Rasenkatz patent. (The equivalent components in the Rasenkatz Claim 1 are the chassis, motor and cover).

The court in fact found that the patent was indeed obvious on several different grounds: it was obvious because it included the Zundapp. Quite separately, it also included the Flymo hovermower. The patent was thus struck from the register held by the Patent Office.

It is interesting to note that such a decision was so clear-cut that it was not appealed. However, it is rare for patent court decisions to go so far. More usually, a patent may be found perfectly valid, but not infringed. Another possibility is that some of the subsidiary claims are found to be invalid and require amendment for the patent to remain valid. Subsidiary claims are usually much less powerful than Claim 1, and are normally inserted to catch those intent on copying but not willing to proceed so far as to infringe the main claim of a patent.

4 Modern invention

The example described in the preceding section, of the development of lawnmowers, provides an insight into the way in which new designs can be protected from copying so as to give inventors or companies a monopoly on their ideas, and an incentive to make and market their product.

Relatively few patents are actually tested in open court. Indeed, many remain either unexploited or unknown, and those that are used in products are usually not challenged by the competition. A strong patent will, in any case, be difficult to challenge either because the prior art is minimal and/or old, or the inventive step is so dramatic as to not make a challenge worthwhile. Court cases can be very expensive to run to a just conclusion, putting individuals in a difficult position unless they can find financial backing.

The Sony Corporation was challenged recently in the UK for infringement of an earlier patent held by an individual. That individual received legal aid to mount his challenge, but eventually lost the case. On the other hand, some creative individuals, like James Dyson for example, have patented widely and succeeded in bringing their design concept to fruition (▼The Dyson vacuum cleaner▲). Dyson in particular has been successful in challenging other companies for infringement of his patents: he won a notable success in the UK in 2000 against Hoover.

▼The Dyson vacuum cleaner▲

The area of household cleaning tools is one which, until very recently, was rather stagnant, showing little or no innovation. The fundamental invention of the cylinder vacuum cleaner was made by Hoover in the 1920s, and had developed relatively little over the years. It was based on an electric motor pulling air through a porous paper bag. The air in turn was pulled into the bag from the surface being cleaned via a flexible pipe, so that the dust was extracted from the surface.

One problem of this device is that the pores of the bag become clogged with very fine dust, so that the velocity of the air stream drops and the cleaning ability of the device falls with time. James Dyson, who had previously developed a garden wheelbarrow with a rubber ball for a tyre and a plastic platform, conceived the idea of a bagless cleaner when observing industrial cyclones for separating powder materials. Using thermoplastics extensively in its construction, his bagless cyclone cleaner has quickly come to prominence in the market (Figure 2.47).

So how does it work? The transparent collection chamber in the centre of the device comprises an outer zone for the coarse dust and an inner one for the fines. The air and dust sucked into the machine from the floor enters the outer chamber first, and forms a vortex in the confined space. The heavier dust drops out and collects here, while the same process in a smaller vortex inside collects the finer material. There is a filter for the finest particles of all. Plastic mouldings are critical to the design because of their low weight and their capacity to be intricately shaped.

A key concept is the transparent central chamber which allows dust collection to be seen by the user.

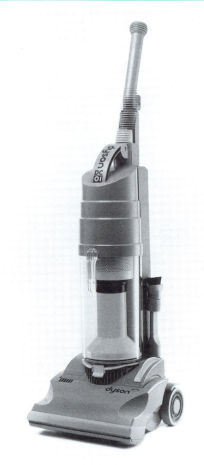

Figure 2.47 The Dyson vacuum cleaner

4.1 New materials in products

As might be expected with a growing class of materials, polymers have been widely used in a large range of new products. The material used in a specific product is not normally named as such in patent claims, so there is potentially a wide range of possible candidates. It also gives the inventor a degree of freedom in selecting the optimum material for the product.

However, choosing a material to some extent determines the manufacturing route which can be used with that material. It may also dictate the optimum design geometry. The Catnic lintel was designed primarily with steel sheet in mind for its construction, and the geometry of the lintel that was patented would not be practical in concrete or wood.

The following case study illustrates the problems which can arise when the original inventive concept was actually developed with only one material in mind. A competitor came along and found that material substitution was possible, and generated a new design that produced certain advantages in the marketplace.

This put the original inventor in a quandary: could the new product be sued for infringement of the concept?

4.2 The wheelie bin

The wheelie bin (Figure 2.48) was invented in Germany, and has now become ubiquitous for disposal of household rubbish. The critical inventive step is the creation of a reinforced lip at the front of the bin which enables it to be hoisted automatically by an hydraulic lift onto a disposal lorry (Figure 2.49). The reinforcement is achieved by bracing of the rim: reinforcing struts can be seen on the bin in Figure 2.48 (see ▼Increasing stiffness▲ on p.86).

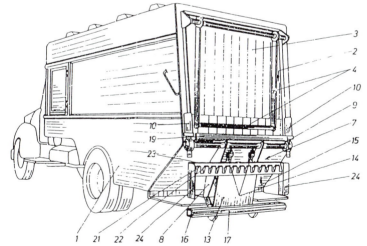

Figure 2.48 Domestic-sized wheelie bin

Figure 2.49 Rear view of a dustbin lorry showing the lifting bar with forks (labelled 22)

The CD-ROM contains the wheelie bin patent (which I will refer to as the 'Schneider patent'). The language in this patent is more complex than in the previous ones we have examined, as it describes in great detail the exact design of the lifting flange.

Have a look at the preamble to the patent, on the first page. You may want to look at the claims on the third page onwards, but do not be worried about the level of detail regarding the flange design. The detailed technical language of structures used in this and the other patents is outside the scope of this course.

▼Increasing stiffness▲

I have already mentioned (in Block 2 Part 1) that polymers have low Young's moduli compared to metals. This means that products made from them will have to find a way to achieve the necessary stiffness.

Exercise 2.6 (revision)

Without changing material, how is it possible to increase the stiffness of a component?

There are ways of giving increased stiffness which do not involve using thicker sections. Indeed, sometimes scaling up part of a product to increase stiffness locally is undesirable, for reasons of aesthetics, or the extra material costs involved.

Using a material with a higher Young's modulus, like steel, for an insert is one possibility, but this introduces an extra assembly stage to the product manufacture, with associated cost.

A favoured option is to build the product with stiffening ribs or struts, to support areas which are likely to deform under load. The triangular struts seen on the bin in Figure 2.48 give such stiffening. Many wooden chairs have some form of bracing between the legs to give the structure extra stiffness. Too much deflection of a part may cause stresses that lead to failure, at a joint, for example.

SAQ 2.12 (Learning outcomes 2.1, 2.3 and 2.6)

What does the patent preamble indicate are the problems of existing refuse bins?

The patent indicates that a plastic material is the preferred material of construction. Part of the design problem with using a thermoplastic material is ensuring that the product is rigid enough to withstand the critical lifting phase. Since the wheelie bin will be full of rubbish, the lifting lip must not deform too much, and the material must also be tough enough to resist cracking. Claim 1 of the patent devotes considerable space to defining this feature in some detail: it is the design of a bin that can be lifted by the rim many times without failure which is the invention in this case.

If we break down Claim 1 as we have done previously, it reads as:

WHAT WE CLAIM IS

1   A container for refuse, comprising

(i)   a body of substantially rectangular cross-section having a cover therefor,

(ii) a depending flange which extends along one side wall of the body in the region of an upper edge of said wall and is stiffened by

(iii) substantially vertically arranged struts connected to the body, and

(iv) a rib which extends substantially parallel with the body between the body and the depending flange; and wherein

(v)   a reinforcing member is provided for said flange, said member extending along and projecting laterally from said flange and being adapted and arranged to serve as an abutment for said container during emptying thereof.

The 'depending flange' is the lifting flange or lip. I have already highlighted the struts on the design. The rib and reinforcing member are shown in Figure 2.50.

There are therefore five essential integers of the refuse bin, all of which must be present in order for the device to function as intended by the patentees. All the integers are concerned with the way in which the lifting flange is stiffened. Remember that polymers have low Young's moduli compared to metals, and in order to make the flange sufficiently stiff, several forms of bracing and reinforcement are provided.

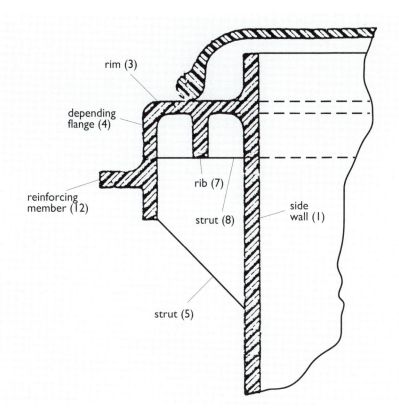

Figure 2.50 Elements of the Schneider lifting rim

4.2.1 Industrial bins

Figure 2.51 accompanying the patent shows just how the metal forks of the lift on the lorry are designed to mate with the cavity below the front flange in order to empty the container (they are shown by the lower set of dotted lines). Another important feature of this claim is that, as again is normal in patents that describe mechanical objects, no dimensions are provided in the claim. In fact (as you may observe just walking down any town street), there is a variety of sizes of such bins, ranging in enclosed volume from about 250 litres (0.25 m^3) up to large industrial bins of 1100 litres (1.1 m^3) total capacity. The former are usually supplied to domestic premises, the latter to shops and businesses.

The large bins are typically about twice as long, on the lifting edge, than the small bins. The length of the lifting edge is increased, but the dimensions of the lifting flange remain the same. This is another example of standardization: constructing the bins like this allows one lorry to handle all the different sizes: a single large bin can be lifted by the same set of forks used to lift two small bins in a single operation.

The Catnic lintel is also a standard product offered in a variety of sizes, depending on the width of the gap to be bridged. However, offering a product in several sizes is not in itself an example of standardization; but where that product has at some stage to mate with another to perform its function, then standards come into play. The lintel, for example, must match not just the aperture to be bridged, but also the cavity in the wall and the brick width, so is a standardized product. It is of course a permanent fixture, while a refuse bin sits free for most of the time. During emptying, however, the cavity under the flange must fit the standard size of comb bar, so the cavity must be standardized for all sizes of refuse bin.

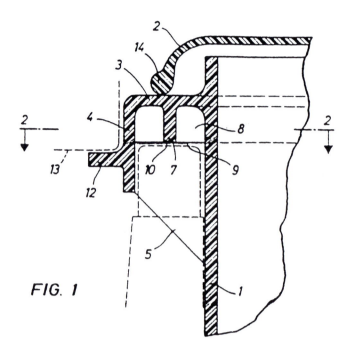

1588932   COMPLETE SPECIFICATION

3 SHEETS   *This drawing is a reproduction of the Original on a reduced scale*

Sheet 1

FIG. 1

Figure 2.51 The design of the Schneider lifting rim

Exercise 2.7

Does the weight carried by the larger bin put extra demands on the flange? Calculate the weight per metre of the flange for a small (250 litres) and a large (1100 litres) bin.

Answer the question by comparing the weight of refuse each bin would hold when full. The density of refuse is about 500 kg m^{-3}. Assume that the flange of the 1100 litre bin is 1.2 m long, and the flange of the 250 litre bin is 0.6 m long.

As the load on the flange for the large bin is more than twice that of a standard domestic bin, extra reinforcement of the lifting flange is needed. The ribs and struts which are a key design detail of such bins provide the extra stiffening where it is needed so as to resist the bending loads imposed during the lift. Many different design options are shown in the Schneider patent in Figures 5 to 9 inclusive. Irrespective of the kind of material used, using ribs and struts is a universal design strategy for controlling product stiffness. On a larger scale, for example, ship hulls made from steel are reinforced internally by providing ribs and stiffeners both laterally and longitudinally. They are welded to the structure during construction.

4.3 Materials of construction

So what is the material of choice for the wheelie bin? When first developed in Germany, they were injection moulded in HDPE (high density polyethylene), a low cost but yet very tough thermoplastic. This makes for a light product capable of withstanding much mechanical abuse, an essential part of the product specification.

The process requires considerable capital investment both in the large moulding machines and the precision tools needed to create the shape. Such machines work continuously, so it is essential to ensure that there is a large enough market to absorb production. In the UK, the use of wheelie bins is mainly decided by local authorities, and their introduction has been a slow, stepwise process. The authority has a large investment, not just in the bins themselves but also in the lorries fitted with the lifting forks. Because we all produce an increasing amount of rubbish, the old rubbish bins could not cope effectively with the volume generated. An intermediate solution involved the use of simple plastic bags, not a good long-term solution because of their low strength. With the advent of recycling schemes, several wheelie bins are provided by some councils for domestic sorting of the rubbish, with separate bins for compostible waste.

However, the large wheeled industrial rubbish container is a product exposed to problems not usually met by the smaller, domestic bin. They are present outside business premises in many public places, and durability can be a problem. Fire can be a specific hazard, especially from fires started by vandals. Polyethylene burns fiercely when the rubbish inside is ignited, so there is a case for an alternative material of construction. Sheet steel is the obvious alternative because it is fire-resistant and incombustible, but how to make a steel bin in the volume needed? Injection moulding cannot be used for steel, and the existing design with extensive ribbing is simply too complex – and hence costly – for extra welding operations.

4.4 The steel bin

During the 1980s, an entrepreneurial company (EH Taylor Ltd.) based in the UK decided to enter the market with a sheet steel 1100 litre wheelie bin. Construction of the basic box shape is not difficult and could be achieved by welding sections together.

But the critical lifting flange needed rethinking. The complex design of the lifting flange for the standard bin is required simply because the product is being made from a polymer, and a structural solution needs to be found for overcoming the problem of low stiffness. Once the decision has been made to change to a different material, a different solution to the lifting problem can be found.

Figure 2.52 The Taylor Continental 1100 litre wheelie bin showing the hollow bar used as the lifting lip

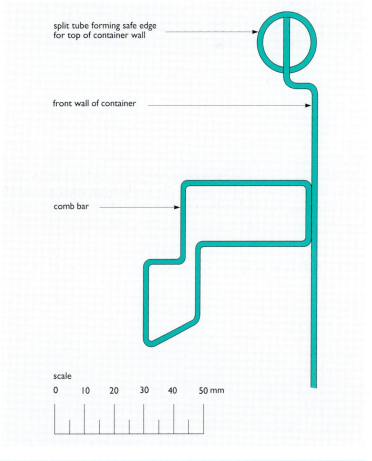

split tube forming safe edge for top of container wall

front wall of container

comb bar

scale

0   10   20   30   40   50 mm

Taylor came up with the idea of using a flange made as a hollow bar welded to the chassis (Figure 2.52). The concept exploits the monocoque idea: like the hovermower case and the Catnic lintel, there is no internal bracing. The flange could be made of the same overall dimensions as the plastic version so that the lorry forks could engage correctly, and the mechanical properties of steel can easily accommodate the lifting loads.

EH Taylor Ltd. went on to capture about half the market for the 1100 litre bins, as you may observe at most recycling centres or in the high street. The steel wheelie bin thus impacted the producers of the plastic bins rather severely.

4.5 Trial

Naturally, the design attracted the attention of the patentee of the plastic equivalent, and a dispute arose which eventually came to trial in the patents County Court in London in 1994. The problem faced by the patentee was simple: was Claim 1 infringed by the steel product? As usual, the legal problem starts with the claim and its technical interpretation by a skilled reader.

SAQ 2.13 (Learning outcome 2.7)

Looking at the flange design in Figure 2.52, refer back to Claim 1 of the Schneider patent. Do you think that the Taylor design infringes Claim 1? (All the essential features must be infringed for the claim to be infringed).

The difficulty the patentee faced in this case was great, because of the narrow way in which the claims had been drafted. This is a perpetual problem for patentees (or the patent agents who usually draft such documents). Draft your claims too wide and you stand the chance of catching not just infringers but also the prior art (in which case your patent is obvious and hence invalid: like the Rasenkatz). Draft it too narrowly, and the claim fails to give you the protection you desire for a complete monopoly; which is what happened in this case. At trial, the patent was found to be valid, despite an attack on its validity, but not infringed. The patent had been written for a thermoplastic injection-moulded product with a specific design, and there was no way a steel flange could fall within the close wording of Claim 1.

The steel bin is a good example where a designer could avoid infringement by changing the design and taking advantage of a different material for construction. Since steel is so much stiffer and stronger than thermoplastic, hollow steel bar would be sufficient for the vertical lifting lip. The Schneider patent specified a lip reinforced by ribs and struts, and Claim 1 was written with these design features as essential integers.

It is interesting to note that the Taylor steel bin was itself patented, and used to pursue alleged infringers of the product! The company has proven the commercial and technical quality of its bin, and competed very successfully in a large market with its plastic equivalent. The domestic market remains safe for the original patentees, but much of its trade in the large bins has been lost to a product which offers that little bit extra in terms of longevity to the user.

In both cases, then, the inventive concept was not the bin, or even the lifting flange, but rather the design of the flange which allowed it to be made with sufficient stiffness in a particular material.

SAQ 2.14 (Learning outcome 2.6)

Referring back to the features of an 'invention' that I defined in Section 1.4.1, discuss why the following are (or are not) inventive products. Make sure that you justify your answers in terms of the three main points.

(a) The Taylor steel bin

(b) The Rasenkatz lawnmower.

5 Conclusions

In our detailed examination of three patents, we have looked at the Catnic lintel, the Black & Decker (Rasenkatz) lawnmower and the Schneider refuse container. They are all products of relatively simple function and shape, and each patent was tested by being used as the basis for court action.

Most patents are not tested anywhere nearly so severely, but are subjected to examination by the Patent Office in the UK, or by the European Patent Office (EPO), if patentees wish for wider recognition of their inventive concept. On a world stage, an inventor will want protection in any market where selling the product is possible. If the invention is revolutionary or quite distinct from the prior art, then this is the route the inventor will take. James Dyson, for example, launched his concept in Japan and the USA before the UK, and has fought many patent actions with infringers and copiers of his basic concept.

For practising engineers, developing new ideas is central to their professional activities, but are there any other ways in which they can protect their innovations?

5.1 'Intellectual property' protection

All new product design work is a form of 'intellectual property' which can be protected from copying by others. For individual inventors, the cost of patenting and fighting infringers can be prohibitive. With companies, the situation is quite different, and they can usually afford the cost of a legal battle, especially if the market is lucrative (such as lawnmowers). But there are other forms of protection.

For example, a company might wish to protect its invention not by patenting but by keeping it secret. This is a dangerous strategy for many products however, because they can be disassembled and recreated, a process known as 'reverse engineering'. Most companies working in a competitive environment do in fact routinely buy their competitors' latest products so as to reverse engineer them, and find out exactly what improvements have been made; presumably with the aim of copying those improvements which could be of benefit to their product, if they are not protected by patent. Design data and 'know-how' fall into the area of 'confidential information', which is one form of intellectual property which can be protected by confidentiality agreements, for example.

Other forms of protection of products include:

registered design (which protects the product's external form and appearance);

copyright in industrial drawings;

design right;

trade marks.

As the name implies, registered designs must be filed with the Patent Office, although the process is much simpler than with patents. Thus several views of the product together with a brief description suffices to define the appearance. Evading a registered design is however, somewhat easier, because it is only the specific *form* shown in the document which is protected: not the function. The general concept of a particular form is not defined, and as long as the copyist makes significant changes so that the consumer is not confused, then he or she generally succeeds.

▼Spare parts for cars▲

One cause célèbre of the 1980s (which has rumbled on and off ever since) is the problem of spare parts for cars, especially common components which need frequent (and often costly) replacement. The problem arose from copyright in the drawings needed to make those parts (at least pre-1989), particularly exhaust systems for British Leyland (BL) cars. The main branded supplier was Unipart, then a subsidiary of BL. Most car owners will know of the problem of corroded exhaust systems, often at a key point such as the junction with the exhaust outlet on the engine. The whole system must be scrapped although other parts may be perfectly adequate.

Then, as now, spare car parts tended to be charged at a premium, and consumers often have no alternative source of supply. However, the technology in producing exhausts is not especially difficult, involving sheet metal fabrication (shaping, welding etc.), and is well within the capability of many small workshops. The system must of course mate with the particular car to which it is fitted.

BL sued one such supplier in the 1970s, charging that the small supplier copied its drawings, or at least the dimensions needed to make the part fit the car. The case ended up in the House of Lords in 1982, and although BL won, the case raised so much concern that the law itself was changed to increase competition.

The point here is that copyright lasts much longer than a patent, being 70 years after the death of the copyright holder, compared with 20 years from the priority date for a patent. The copyright holder was effectively given a monopoly for relatively low level technology lasting much longer than for a more advanced, expensive and short-term patent.

The result was the Copyright, Design and Patents Act of 1988, which replaced copyright in industrial designs with 'design right'. This entirely new 'intellectual property' right arises automatically, just like copyright, but is of much shorter duration, lasting only 10 years from selling of a new product, or 15 years from the creation of a design whether sold or not. Parts of that product which 'must mate' with another are specifically excluded, as are 'methods or principles of construction'. Thus the attachment dimensions for a car exhaust system, such as the key connection to the engine and parts of the body, are excluded from any protection at all. It allows independent companies to make and supply such spare parts, so saving costs for the consumer. Many such so-called 'generic' components are now readily available to car users as a direct result of the new policy, although small manufacturers often find it difficult to keep up with all new models introduced into the market.

Copyright and design right are two forms of intellectual property protection which are unregistered, and automatic, so the cost of protection is much lower than with patents or registered designs. Copyright is a familiar form of protection to authors, and recording or performing artists, and an extension of the same idea also covers product design drawings (but only if produced before 1988/9). After this date, protection comes in the form of design right, a kind of protection unique to the UK. Like copyright, the protection is automatic as long as the originator can prove originality, and the design or its detail do not involve a principle or method of construction. Specific design details such as a new configuration of a set of gear wheels, or a new way of screwing parts together, can also be protected by this right.

There is an important exception to this right: those parts which mate with other, standard parts cannot be protected, an exception which was created by the problems of designing ▼Spare parts for cars▲: the problems of independents making exhaust pipes, for example.

The period of protection is shorter than with patents (about 15 compared with 20 years), and there are several problems in interpretation of the statute which can only be resolved by court actions. As we saw with the Catnic case, the law in this specialized area is determined not just by the statute drafted by Parliament, but also by the interpretation of the statute in case law. Although a flexible and almost universal tool of expression, language presents many ambiguities in attempting exact definition of engineered products.

5.2 The range of mechanical invention

The case studies we have looked at involved technical effects which were entirely mechanical in design, but of course, they represent only a tiny part of the field of invention. Contrary to popular belief, invention does not always involve making complicated and involved pieces of machinery. The idea of

▼The Workmate▲

This product was invented by Ron Hickman in 1968, when faced with the problem of planing and modifying large workpieces such as door and window frames. He had moved into an empty house and intended to make his own fitted furniture. Large pieces of wood are unwieldy for a single person to work on, and a single small vice is not capable of holding them securely, only being able to grip a small part of the structure. Working at a distance from the gripped section can put high stresses on the vice by a lever action, and the workpiece can move and so become even more unstable.

Hickman tackled the problem by devising a sawhorse fitted with a large vice, which could hold the piece in a stable position on which it could then be worked. Several prototypes later, he had developed the idea of two vices working independently at either end of the table but connected by a single pair of jaws. They effectively converted the whole of the table top into a giant vice, so enabling very large pieces of timber to be held and worked more easily. Since the two vices were also independent of one another, tapering samples could also be gripped in a stable way. The device was finally awarded a full patent in 1972, following a provisional application in 1968. The patent was licensed exclusively to Black & Decker in 1972, with production climbing from about 100 per month initially to about 68 000 in 1976. The preferred embodiment of the concept is shown in Figure 2.53, where the table is supported by a collapsible metal frame. This innovative feature allows easy transport and storage in a small space (such as hung on a wall), yet still providing strong and reliable support for the top working surface.

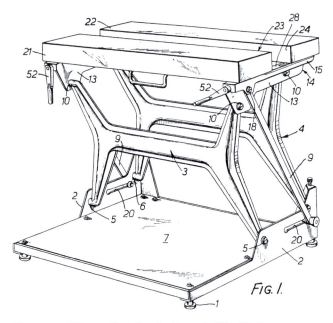

Figure 2.53 The 'preferred embodiment' of the Workmate

the cyclone applied to a domestic cleaner, for example, is not especially difficult to appreciate once the basic idea is explained. The engineering principle was well-known, but it did take an inventive step to apply that idea to solving a quite different problem. Similarly, the idea of a temporary folding workbench with an adjustable vice-like worktop uses well known mechanical principles to achieve its effect. However, it addresses a problem faced by many consumers, so the invention of ▼The Workmate▲ proved an instant success when it was launched by Ron Hickman.

Another area of vigorous inventive activity is medical devices. For example, heart surgery has been transformed in its life-saving effects by the invention of the balloon catheter for removal of fatty deposits in arteries. A more recent development has been the arterial stent, which acts as a brace within the artery after the operation (see ▼Heart surgery, balloon catheters and stents▲).

5.3 Other areas of invention

In a section of this length, it is impossible to cover the many areas of active innovation, although some obvious areas include:

electrical engineering and electronics;

pharmaceuticals;

polymer technology;

computer hardware and software.

▼Heart surgery, balloon catheters and stents▲

One of the most remarkable advances in recent years has been the development of heart angioplasty procedures for diseased arteries. This invasive surgery involves feeding a catheter into a major artery (usually in the groin). A balloon is fitted over the catheter, so that when the catheter has been fed to the diseased part of the artery, it can be inflated and so widen the constriction at that point. Narrowing of the arteries is caused by a build-up of fatty tissue (arteriosclerosis). The original invention of the balloon catheter was made by a Swiss surgeon, Andreas Gruntzig, who fed the balloon over a guide wire through the catheter. One problem with the operation is that the fatty tissue build-up recurs (restonosis). To prevent this happening, so-called stents were developed by several surgeons. The stent is a perforated metal sleeve which fits over the balloon, so that when the balloon is inflated, the stent expands and remains in place owing to plastic deformation in the metal (Figure 2.54).

One of the original pioneers, a US surgeon, Dr Julio Palmaz, patented his design of stent, which he then used in infringement proceedings against one manufacturer (Cook Inc.) in the state of Indiana. The issue came to be tried first in the UK however, when Boston Scientific asked the courts to revoke Palmaz's European and UK patents. The trial took place in 1998. The outcome was that the scope of the original Palmaz patents was substantially curtailed.

This area of research has produced many different stent designs from different surgeons, who are now free to use them without the chance of infringement proceedings for what has become a life-saving operation for many heart attack victims.

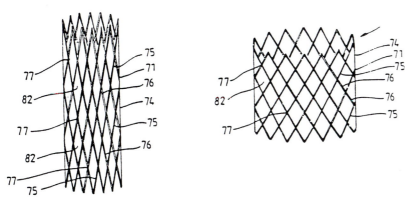

Figure 2.54 Original stent at left, expanded stent at right (from Palmaz patent)

Even within areas such as electrical engineering, many devices still work mechanically, such as residual current devices (RCDs) which trip when a leak is detected in an electrical circuit. While the earliest devices for protecting against damaging surges of current in circuit were simply fine wires (the fuse, invented by Edison in a patent of 1880, as part of his development of the electric-light bulb), they cannot prevent electrocution because they act too slowly. As the current rises, the temperature of the fuse wire rises to incandescence and finally melts to break the circuit: this takes time, more time than is safe to prevent electrocution. Circuit breakers are also used widely to protect equipment but are again too slow to prevent electrocution. The invention of a life-saving circuit breaker (RCD) was only surprisingly recently, being based on a Westinghouse patent from the 1980s. Such devices are available to protect external circuits such as the wire leads to lawnmowers.

Advances in chemical synthesis have, since the 18th century, been among the most active of subjects, and that progress continues in pharmaceuticals and polymers. The synthesis of new drugs is now very rapid, especially with detailed understanding of the role of DNA in living things. New polymers continue to be made, with many stunning discoveries of the last thirty years, such as aramid and polyethylene ▼High performance fibres▲

▼High performance fibres▲

The goal of highly-oriented fibres has been pursued for many years, ever since theoretical chemists were able to show that the theoretical strength and stiffness of polymer chains approached if not exceeded that of the strongest and stiffest steels. In order to exploit these properties, though, the fibres produced need to be well-aligned with the carbon backbone of the polymer chains.

The first experiments (harking back to Edison and Swan's attempts to use the material in electric-lamp bulbs) in making carbon fibre started with a polymer precursor, polyacrylonitrile fibre (PAN). It is commonly known as 'acrylic fibre' and finds wide application for textile products. The Royal Aircraft Establishment (a UK defence research establishment) at Farnborough formed, in the mid-1950s, a higher strength fibre from PAN by heating (so-called 'black Orlon'), and then converted that material to carbon fibre by stretching and pulling the fibre bundles.

The resulting material, when combined with a matrix resin to form strong composite materials, has since found wide application, used in demanding structural roles (racing car bodies, aircraft bodies etc.). Composites are an important class of material, where mixtures of materials are used to exploit beneficial properties of the different parts of the mixture. So carbon fibres might be bound together in a polymer matrix, or a metal might be reinforced with stiff ceramic particles.

However, carbon fibre is expensive to make in batches, and chemists have sought to make a continuous fibre by other routes. The principal breakthrough came in 1968, when a new fibre with an *aromatic* backbone chain (Figure 2.55: aromatic is a chemical term referring to the hexagonal rings of carbon atoms in the chain backbone) was synthesized by DuPont in the USA. It was developed into a fibre which could be made in continuous form, and subsequently woven into fabrics, or products like rope. Known by the tradename Kevlar, it has become widely used not just in composite form but also as a reinforcing fibre for high-performance materials. It also has fire-resistant properties, which make it ideal for fire-resistant products. A close relative, Nomex fibre, is used for firefighters' garments.

In 1988 the company DSM developed a high tensile fibre known as Dyneema. Made from very high molecular mass polyethylene by a process known as gel-spinning, it is widely used for ship-mooring ropes owing to its strength and low density (lower than water), where it has displaced steel cables.

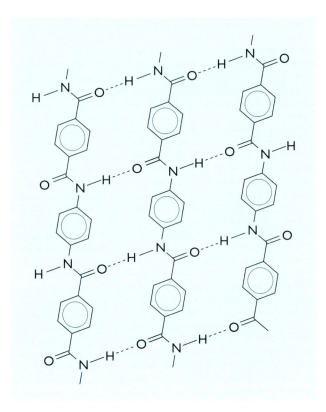

Figure 2.55 Structure of Kevlar fibre. The figure shows three polymer chains running from top to bottom (you saw pictures like this that represent chemical structures in Block 1 – Figure 2.14, page 82). Here we show three chains. There are weak chemical links between each chain (the dotted lines between the O (representing oxygen) and H (representing hydrogen) atoms). The hexagon symbols are a shorthand for a 'ring' of six carbon atoms, each joined to a hydrogen (as was drawn out in full in Block 1 Figure 2.14(c)). You are not expected to remember these details! Just note that Kevlar's properties depend on the chains being well-aligned in the fibres.

5.4 The sources of invention

So why are such advances important to engineers? After all, many recent and spectacular inventions have been led by scientific discovery rather than the guided empiricism of people like Edison. The answer is more complex than we might think initially. Edison in fact pioneered areas of research which were later followed by scientists, such as the way metal filaments in bulbs broke down at high temperatures. Much of his work, assiduously recorded in his notebooks in 1879–80, shows an attention to the fine detail of breakdown, especially in his microscopic examination of his filaments after failure. He recorded details such as grain (or crystallite) structure in the metals (especially platinum) and how it influenced failure. Such work laid the basis for the later adoption of tungsten metal as the principal filament material for light bulbs, in place of the initially successful carbon filaments. The idea of the electric light bulb also led to many new inventions for supply of the power needed for his bulbs, such as fuses, distribution networks, bulbs and sockets, dynamos, vacuum pumps, and new insulators. Much of his inventive work relied on knowledge developed in other fields (such as telegraphy). New materials or new ways of manipulating well-known materials also lay at the heart of his activity, with carbon eventually being adopted for his first bulbs after exhaustive trialling of many other materials in his labs.

Much of this is still true today, with innovation in materials technology leading to development of new devices which exploit the new properties offered. Thus composites technology owes its great advances in applications ranging from bike frames to aircraft structures to the invention of carbon fibre, aramid and other high performance fibres. The story could be repeated several times.

Serendipity also plays an important role in invention: the chance and unexpected discovery has been critical to many inventions, such as steel by Bessemer. Blowing air through the molten iron caused an explosive reaction which raised the temperature and burned off the excess carbon. At the time he thought he had made a mistake, but the iron produced was almost free of carbon, and the explosive reaction was thus the key to the new process. Many polymers were found in a similar way, when reactions produced strange solids not initially expected. Both PTFE and polyethylene were discovered in this way. Similarly, the vulcanization of rubber was a serendipitous discovery by Goodyear, when he found accidentally that sulphur could enhance the physical properties of rubber.

5.4.1 Patents and invention

Inventors try to find solutions to hitherto unsolved problems, but how do they know whether or not someone has beaten them to it? They will turn first of all to the patents already granted in their particular field of activity. A patent search can be performed very easily now on the internet via several official web sites. The patent databases are by far the largest single repository of technological information in the world, so access to this source of information should prevent the problem that Dunlop faced with his rubber tyre!

Recognition of a problem is just the first step in the path of discovery which any inventor then faces in trying to solve that problem. When a solution has been found or achieved by any one of a number of routes, the inventor creates a prototype to which improvements can be made. The end result should be patented to protect the inventive concept, which inevitably means articulating a description of the preferred embodiment with the crucial claims for the idea. Claim 1 in particular will describe the breadth of the concept, and to what extent it is distinguished from the prior art in that field of activity. Too close, and it will run the risk of including prior art and so be obvious, but if too narrow, will not catch any imitators. Once the patent is published, it becomes public knowledge, and thus accessible to other inventors and manufacturers who might wish to design around the idea, and so exploit any market. If the invention breaks entirely new ground, it will be difficult to copy it exactly, but competitors may well improve or change the concept and gain an edge in the market (as we saw with the refuse bin).

Design, innovation and improvement lie at the heart of engineering, and understanding ways in which those innovations can be protected is a necessary prerequisite for successful engineering and product design. When all is said and done, better products enhance our lives and contribute significantly to the well being of nations.

6 Learning outcomes

At the end of this block, you should be able to:

2.1 Identify the problem which an inventive concept seeks to solve.

2.2 Given appropriate information, identify the inventive concept in a simple device and describe how that concept is realized in a product.

2.3 Describe the previous practice (or 'prior art') in a field before a new invention was available.

2.4 Describe the function(s) of a simple product and the materials properties needed to meet those requirements.

2.5 Indicate the criteria which need to be met for a new product to be regarded as inventive.

2.6 Read a patent document and appreciate the problems addressed, how the inventive concept functions and how it addresses those problems.

2.7 Critically examine features of a specific product against the technical claims of a patent, and understand the way patent claims can protect an inventive concept.

2.8 Understand that shaped steel beams have different values of stiffness, as measured by their second moment of area I, and that this affects how they deflect under load.

Answers to exercises

Exercise 2.1

The volume of the beam is given by the product (the multiplication) of the three dimensions, so is

$$1.20 \times 0.15 \times 0.09 = 0.0162 \text{ m}^3$$

(a) The density of the oak beam is 1200 kg/m^3,
so the mass = 1200 kg/m^3 $\times$ 0.0162 m^3 = **19.4 kg**

(b) the density of the concrete beam is 2400 kg/m^3,
so the mass = 2400 kg/m^3 $\times$ 0.0162 m^3 = **38.8 kg**

(c) the density of steel is 7800 kg/m^3,
so the mass = 7800 kg/m^3 $\times$ m^3 0.0162 = **126.4 kg**

Exercise 2.2

The cross-sectional areas of the three beams are:

Square cross-section:

$$B \times W = 0.005 \text{ m}^2$$

The hollow beam and the I-beam have the same areas:

$$B \times W - b \times w$$

$$= 0.005 \text{ m}^2 - 0.0036 \text{ m}^2$$

$$= 0.0014 \text{ m}^2$$

As the length of the beams is one metre, the volumes are simply 0.005 m^3 for the square cross-section beam, and 0.0014 m^3 for the other two beams.

Multiplying this volume by the density gives the answer that the square beam will have a mass of 39 kg, and the other two 10.9 kg, showing that there is a significant weight saving in using the hollow or shaped beams.

Exercise 2.3

(a) The prior art mentioned in the preamble to the description includes lintels made from heavy baulks of timber and heavy gauge metal.

(b) The main problem of the prior art lintels is that they are heavy.

(c) Among the main aims of the invention are to provide lintels which are:

light in weight,

inexpensive, and

easy to handle.

Exercise 2.4

(a) The preferred material is sheet steel with a closely adhering coating of zinc to prevent corrosion (i.e., galvanized steel).

(b) The sheet steel should conform to BS 2989[1].

(c) Other materials which could be used in the lintel include, for example,

extruded aluminium or

plastics reinforced with carbon fibre or

metal coated with a protective layer of plastics material.

[1] In fact, this standard has since been superseded by BS EN 10143.

Exercise 2.5

In this case, the angle between F_a and F_v is $(90 - \theta)$ – see Figure 2.27(b), so the answer is:

$$F_v = F_a \cos(90 - \theta)$$

Exercise 2.6

The stiffness of a component can be changed by increasing the amount of material present: by making a beam thicker, for example.

Exercise 2.7

The large bin has a volume capacity of 1100 litres (1.1 m^3) compared with just 250 litres (0.25 m^3) for the small domestic bin.

The mass of rubbish will be: $1.1 \text{ m}^3 \times 500 \text{ kg m}^{-3} = 550 \text{ kg}$ for the large bin.

$0.25 \text{ m}^3 \times 500 \text{ kg m}^{-3} = 125 \text{ kg}$ for the small bin.

The load per unit length of bin flange is then:

1100 litre bin:

load supported = 550 kg/1.2 m

= **458 kg** per metre of flange

250 litre bin:

load supported = 125 kg/0.6 m

= **208 kg** per metre of flange

The loading per unit length of flange on the large bin is thus over twice that of the small bin, so extra support is probably needed in the flange of the large bin. It could be achieved with extra ribs and struts above and below the flange, and by thickening the wall of the product.

Answers to self-assessment questions

SAQ 2.1

(a) The problem which led to the flame safety lamp was the widespread occurrence of methane explosions in collieries. They were caused by ignition of the methane by open flame lamps used for illumination in the pits.

(b) The single key idea which led to a solution of the problem was the observation that iron gauze of fine enough mesh size would prevent a methane flame penetrating through it.

(c) The idea was used by Davy to construct a safety lamp by enclosing a simple flame lamp with a cylindrical gauze capped by another piece of gauze of the same mesh size (~1 mm). Any methane gas in the colliery air would burn safely within the lamp, giving a blue cone above the bright part of the flame. It would be prevented from reaching the outer colliery atmosphere by the gauze.

(d) The main practical problem which arose as a result of use of the design in working collieries concerned rusting of the gauze. Moisture could attack and corrode the wires, and if just one such wire rusted through, the lamp became unsafe, and a flame could pass through and cause an explosion. The rusting problem was solved by providing more than one enveloping gauze, so that if one rusted through, another was there to provide protection. The flame could also be protected by a glass cylinder below the set of gauzes.

SAQ 2.2

(a) The critical problem in making an incandescent electric lamp lay in finding a material which would resist burning in air at the high temperatures created when an electric current flowed through the filament.

(b) The previous practices (the 'prior art') in the field of illuminating lights before 1879 included

1 open flame lamps or candles

2 flame lamps with a mantle

3 lime lights, and

4 carbon arcs.

Open flame lamps provided very little illumination, but it could be increased by providing a mantle around the flame. These were composed of a woven structure impregnated with chemicals which increase the illumination level. One such salt was lime, which created an intense light when heated by an open flame. The carbon-arc lamp consisted of two electrodes supplied with electricity which when close enough, allowed a continuous spark to travel between the electrodes. Although the arc provided a very bright light, the carbon rods burned away rapidly.

(c) The material choice for a filament material for an electric lamp is important because the following properties are needed:

1 very high melting point

2 electrically conducting

3 capable of being made into a fine filament.

SAQ 2.3

(a) The main function of a ladder is to provide access to the upper parts of a wall or building which are normally inaccessible. In normal use it will be leant at an angle to a wall, so that the user can climb to a given position and gain that access. It must support the user in a stable and safe way, so the materials of construction must resist deformation under the weight of the user (plus any additional load carried by the user, such as paint or tools). The ladder is normally used at an angle of about 75 degrees to the horizontal (as shown by a warning notice posted on all new ladders), and the ladder will normally be fitted with anti-slip devices. Since it is certainly used externally, and may be left outside for some time, it should be resistant to the weather (rain, frost, sunlight etc.). It should also be light enough in weight to be moved easily.

(b) The material used for the ladder should be strong, so that it does not fail under the user's weight. It should also be stiff, so that it does not flex unduly during use. It should be resistant to degradation if left exposed to the elements.

(c) The main materials of construction include wood and aluminium for both the stiles and rungs. Wooden ladders usually rely on unprotected feet and tips; aluminium ladders are normally equipped with rubber feet and plastic tips to prevent slip during use. While both wood and aluminium are intrinsically stiff enough to provide stable ladders (they have reasonably high Young's moduli), problems could arise if the thickness of the component parts was too low. Both materials offer acceptable resistance to deformation caused by the user, but there are differences in environmental resistance. Wood in particular is susceptible to rot from fungi, algae and bacteria, which can cause major structural weakness in the long term. Aluminium is much less susceptible to such degradation.

SAQ 2.4

Davy's lamp design is inventive because:

1   It was novel: no one had made such a design before.

2   Presumably the use of the gauze was not obvious at the time, or it would have been used previously!

3   The concept was applied successfully to making a product, so it was clearly capable of manufacture.

SAQ 2.5

(a) For the square section:

$I = 0.05 \text{ m} \times (0.1 \text{ m})^3 \div 12 = 4.2 \times 10^{-6} \text{ m}^4$

(Note the units of I are m^4, reflecting the multiplication of lengths in the formula to calculate it.)

For the hollow section:

$I = (0.05 \text{ m} \times (0.1 \text{ m})^3 \div 12) - (0.04 \text{ m} \times (0.09 \text{ m})^3 \div 12) = 1.7 \times 10^{-6} \text{ m}^4$

Although the hollow beam is around a quarter of the weight of the solid beam, its value of I is only reduced by just over half. This shows the benefits of removing some of the material from the centre.

(b) Using the formula

$$\Delta = \frac{F L^3}{48 E I}$$

for the solid section:

$\Delta = (1000 \text{ N} \times 1 \text{ m}^3) \div (48 \times 210 \times 10^9 \text{ N m}^{-2} \times 4.2 \times 10^{-6} \text{ m}^4)$

$= 2.36 \times 10^{-5} \text{ m}$

And for the hollow section:

$$\Delta = (1000 \, \text{N} \times 1 \, \text{m}^3) \div (48 \times 210 \times 10^9 \, \text{N m}^{-2} \times 1.7 \times 10^{-6} \, \text{m}^4)$$

$$= 5.84 \times 10^{-5} \, \text{m}$$

So in each case the deflection is small. From this result, shaped beams look like a good idea.

SAQ 2.6

The four essential integers of the Catnic patent were

(i)   a top horizontal plate

(ii) a lower horizontal plate

(iii) a front support and

(iv) a rear support plate.

The Cougar design fails to meet these criteria because it possesses neither top nor base horizontal plates. The top of the Cougar lintel is a simple corner, and the base of the product is hollow. Thus the product clearly falls outside the boundaries of the original Catnic patent. Even if a lower plate were welded onto the structure it would still fail to possess a top horizontal plate. Since it falls outside Claim 1, it also falls outside all the other claims of the patent because they are all narrower than Claim 1.

Therefore, despite being a good product, the Cougar lintel is not covered by the original patent.

SAQ 2.7

(a) The mowers are of course all designed to cut grass, but they achieve it in quite different ways. The cylinder mower has been around for many years, and is probably the oldest of the three general types. Larger models may be seen on ornamental greens, cricket pitches and so on. It achieves its function by means of a set of blades set in a helical path on the outer surface of a metal cylindrical grid. The grid is rotated rapidly and is usually powered by an internal combustion engine. The larger machines have rollers and may also have a seat placed over the roller, so that the user rides the machine when mowing. The height of cut is controllable by changing the height of the rotating blades. Smaller versions are available (hand or machine powered) for small gardens, but are relatively expensive. This may reflect the precision engineering needed for their construction, especially of the rather complex cylinder itself, but also for the drive mechanism.

Hovermowers have been available for a number of years, at least since the 1970s, and provide the same basic function as a rotary mower but in a quite different way. They cut grass using a much simpler rotating blade fixed under a plastic hood, and connected directly to the electric motor. The device floats on an air blast, so the height of cut cannot be controlled at all, but the simplicity of the machine to manufacture means it is relatively cheap to buy.

The wheeled rotary mower cuts the grass by means of a horizontal blade which may be powered by electricity or a small petrol engine. The height of cut may be varied at will simply by adjustment of the position of the wheels. The quality of the cut is variable depending on the initial height of the grass and coarseness of the grass, whether there is substantial weed growth and the slope or roughness of the land. It also depends on the quality of the edge of the blade, which is normally near the tips of the blade. The cut grass may be collected at the rear or side by a duct which connects to the main chamber in which the blade rotates.

Strimmers have been available since the 1980s. Their main function is as a tool for trimming grass edges and even small areas of lawn by cutting it through the action of a rotating plastic filament. The filament is situated on a spool at the end of a pole held by the user's hand, and they may be powered electrically or by an internal combustion motor.

(b) The problem which the inventors of the first rotary mowing machines addressed was presumably the need to cut grass in a systematic and efficient way. Presumably, it was performed by hand using scythes, or perhaps by allowing animals to graze the lawn before the invention of the motor. So the mechanization of the mowing process is the innovation here.

For the hovermower, the problem addressed by the inventor is presumably the high cost of cylinder mowers. Because it has no wheels, the product can cut grass on rough ground very easily. So the innovation in this case is a combination of cheapness and ease of use.

The general problem addressed by designers of rotary mowers is to provide extra control over such features as height of cut and collection of cut grass at a reasonable cost. They are a sort of 'hybrid' design, and the innovation here is combining the flat blade of the hovermower with the stability and control of the rotary mower.

For the strimmer, the problem addressed is the limitation of conventional mowers to cut edges and awkward areas of grass.

SAQ 2.8

(a) The hood fulfils the following practical functions when being used to mow grass:

 1 directs the air blast efficiently onto the ground,
 2 protects the user from the rotating blade,
 3 protects the user from objects thrown up by the moving blade,
 4 supports the motor,
 5 rides the air cushion,
 6 moves into the unmown grass.

You may also have added other functions performed by the hood: for example, the need for transmitting forces from the handle during use for steering, and supporting the weight of the mower if it is hung from the handle during storage. When in use, the handle is at a steep angle to the plane of the hood, but the handle folds flat against the hood when being stored against a garage wall, for example. This function requires some kind of fixture in the hood which will mate with another on the handle to keep them in a stable position to one another. Another important attribute of the hood could include wear resistance and a low-friction inner surface to inhibit grass sticking to the underside. Other attributes of the hood which are important include low raw material cost and ease of manufacture, and it must last for a long time outside.

(b) The property requirements to fulfil these seven functions include rigidity (1, 4, 5 and 6), toughness (2, 3, 4 and 6), low weight (5 and 6), and resistance to corrosion. The additional need for fixtures requires toughness and rigidity of the material of the hood and fixture.

SAQ 2.9

(a) The prior art mentioned in the preamble to the description of the invention mentions 'conventional' rotary lawnmowers, where their structure is described in some detail. The principal component parts are the upper and lower housing, the former also being called the hood. Wheels are mounted on the outside of the lower housing.

(b) The main problems of conventional mowers include:

 (i) uneven lower surface next to the blade so that grass clippings are caught, making it difficult to clean,

(ii) walls and trees cannot be approached closely because externally mounted wheels interfere, and

(iii) the wheels can be jammed in bushes and branches.

SAQ 2.10

(i) The chassis of the RE 42 (Figure 2.42) encloses the rotating blade, so a substantial portion is in the plane of the blade. At the front of the mower, there is an upturned edge which mates with a cover. The rear end is quite different, however. The edge actually points down, and certainly does not mate with the cover (or anything else). This argues against calling it a 'dish-like body'.

(ii) The motor is fitted with an output drive shaft having a free end, and it appears to be mounted on the chassis. The free end penetrates the bottom wall for carrying the blade, so the model meets this particular integer.

(iii) The RE 42 possesses a cover which mates with the chassis, but not at a 'peripheral sidewall' of the chassis.

The conclusion is that the Flymo RE 42 does not infringe two of the essential integers ((i) and (iii)) of Claim 1 of the Rasenkatz patent.

The Zundapp mower from 1969 can be compared in exactly the same way with Claim 1 of Rasenkatz:

(i) It possesses a lower chassis with a peripheral sidewall extending upwards to mate with the cover. It has a substantially continuous bottom wall, and a portion is in the plane of the cutting blade. The peripheral sidewall points down at the sides (side elevation of Figure 2.43) however.

(ii) It also possesses a motor with a free end penetrating the chassis and connected to the cutting blade.

(iii) The cover is coextensive with the uppermost edge of the chassis, and engages with it to define a housing for the motor.

This model seems to possess at least two of the essential integers of Rasenkatz, (ii) and (iii). The chassis does not have a continuous upward sidewall at the sides, but it is only this point which apparently brings it outside Claim 1.

SAQ 2.11

(i) The hood of the hovermower possesses an upside-down saucer-like shape with an outer edge which is upturned. However, it also has another wall extending upwards near the motor. The bottom wall is substantially continuous, and a part is in the plane of the cutting blade.

(ii) The hovermower has a motor which penetrates the hood and has a free end to which the blade is attached.

(iii) The cowl of the hovermower seems to be 'coextensive' with the upper wall of the hood surrounding the motor, and is attached to the hood, so may be said to be 'engageable with' the hood and its upper wall.

The hovermower thus seems to fall within the widest meaning of Claim 1, the weakest point in the argument being the nature of the meeting between the cover and the chassis (i.e., cowl and hood). However, it could easily be designed so that they met, and such a choice would be a design option at the time the original hovermower was made. The inner edge of the hood is not an outer peripheral sidewall, but again, the cowl could be extended to meet the edges of the hood. Indeed, this is just what has been done in the latest designs of hovermower shown in Figure 2.37. The final point is that the hovermower has no wheels.

Thus the hovermower could easily be said to infringe Claim 1 of the Rasenkatz patent, giving the claim the widest possible interpretation.

SAQ 2.12

Existing refuse bins suffer considerable damage at the crucial lifting lip owing to the high strains and impact loads on the lip during the lifting phase of emptying. They are also subjected to widely varying temperatures which lead to deformation and dislocation of the flange, so that they become, in some cases, unusable. One piece of prior art is the German Gebrauchsmuster 76 11 603 where the lifting lip is reinforced with a metal bar between the lifting flange and the body of the container. It makes manufacture more difficult and increases the cost of the bin, the flange is still not adequately protected against deformation or bowing, and 'it does not aid the abutment surface'.

SAQ 2.13

The five essential integers from Claim 1 of Schneider are:

1   rectangular body with lid,
2   depending flange,
3   vertical reinforcing struts,
4   single rib under the flange,
5   external rib for abutment.

All are essential to the invention, so all must be present on the Taylor bin if it is to be judged to infringe the patent.

The diagram (Figure 2.52) shows that the Taylor bin satisfies 1 and 2. It also has welded struts, so satisfies 3. The external part of the flange could be interpreted as an external rib (despite the fact that it is hollow), so integer 5 might be satisfied. However, there is no single rib under the flange, so the Taylor bin apparently falls outside the scope of Claim 1.

SAQ 2.14

Here is my interpretation. You may have made different decisions!

The Taylor bin:

1   Steel bins have been around for a long time, so presumably the novelty lies in a bin that can be lifted automatically by a lorry with appropriate forks.
2   The non-obvious part of the design in this case is the flange by which the bin is lifted. Presumably the design is non-obvious, as the application was new. (The design could, of course, be shown to be obvious if a similar flange was used for lifting purposes in another application).
3   Clearly the bin is capable of manufacture.

It looks as though the bin does count as an invention for our purposes, and certainly for a patent.

The Rasenkatz:

1   As mowers were well-known at the patent date, it is the shape and construction of the mower which is presumably novel.
2   It seems difficult to say that the shape of a mower is 'non-obvious'. However, this is tied in to the way in which it is manufactured, and there are many possibilities for manufacturing. However, as we saw, the mower form in this case was judged to be obvious to a designer working in the field.
3   Again, the mower is capable of manufacture.

The 'obviousness' of the design means that it is non-inventive.

Acknowledgements

Grateful acknowledgement is made to the following sources for permission to reproduce material within this block:

Part 1

Figure 1.3: © AO Safety Products, Watford, England; *Figure 1.13:* Courtesy of P. R. Lewis and G. W. Weidmann.

Extracts from British Standards are reproduced with permission from BSI under licence number 2000SK/0579. British Standards are available from BSI Customer Services, 389 Chiswick High Road, London W4 4AL, United Kingdom (Tel UK 020 8996 9001).

Part 2

Figure 2.1: Singer, C., Holmyard, E. J., Hall, A. R., and Williams, T. I. A. *A History of Technology*, Clarendon Press 1954–8, by permission of Oxford University Press; *Figure 2.6*: James, P. and Thorpe N. (1995), *Ancient Inventions*, Michael O'Mara Books Ltd.; *Figure 2.7*: © National Railway Museum/Science and Society Picture Library; *Figures 2.8, 2.12 and 2.30*: © The Science Museum/Science & Society Picture Library; *Figure 2.9*: Courtesy of the Ironbridge Gorge Museum Trust; *Figure 2.10*: © Dunlop Tyres Limited; *Figure 2.13*: Coulson, J., Carr, C. T., Hutchinson and Eagle, D. (eds.), *The Oxford Illustrated Dictionary*, 2nd edition 1975, Oxford University Press. Bay Books Pty Ltd., (Imprint of Murdoch Books); *Figure 2.19*: © Travel Ink, Derek Allan; *Figures 2.21, 2.25, 2.26, 2.31 and 2.32*: Courtesy of Catnic, part of the Corus Group; *Figure 2.27*: *Reports on Patent Cases*, Vol. 9, 1982, Butterworth Heinemann. The Patent Office. © Crown Copyright is reproduced with the permission of Her Majesty's Stationery Office; *Figure 2.29*: © David Lee Photography Ltd, Humber on Bourton; *Figure 2.34*: Courtesy of Atco-Qualcast Ltd (Bosch Group); *Figures 2.35, 2.37, 2.42, 2.43 and 2.46*: © Flymo; *Figures 2.39, 2.40, 2.49, 2.50, 2.51, 2.53 and 2.54*: Extracts from Patent Specifications, © Crown Copyright is reproduced with the permission of Her Majesty's Stationery Office; *Figure 2.41*: Extract from US Patent 4194345; *Figures 2.38, 2.44, 2.45 and 2.52*:
© Peter Lewis; *Figure 2.47*: Dyson Appliances Ltd; *Figure 2.48*: Courtesy of Otto UK Ltd.

Thanks to the T839 course team for providing material from Block 4 (Intellectual Property Matters).

Every effort has been made to contact copyright owners. If any have been inadvertently overlooked, the publishers will be pleased to make the necessary arrangements at the earliest opportunity.